AF389157

Intelligent Systems Supporting the Use of Energy Systems and Other Complex Technical Objects, Modeling, Testing and Analysis of Their Reliability in the Operation Process

Intelligent Systems Supporting the Use of Energy Systems and Other Complex Technical Objects, Modeling, Testing and Analysis of Their Reliability in the Operation Process

Editor

Stanisław Duer

MDPI • Basel • Beijing • Wuhan • Barcelona • Belgrade • Manchester • Tokyo • Cluj • Tianjin

Editor
Stanisław Duer
Technical University of Koszalin
Poland

Editorial Office
MDPI
St. Alban-Anlage 66
4052 Basel, Switzerland

This is a reprint of articles from the Special Issue published online in the open access journal *Energies* (ISSN 1996-1073) (available at: https://www.mdpi.com/journal/energies/special_issues/ Intelligent_Systems_Supporting_the_Use_of_Energy_Systems).

For citation purposes, cite each article independently as indicated on the article page online and as indicated below:

LastName, A.A.; LastName, B.B.; LastName, C.C. Article Title. *Journal Name* **Year**, *Volume Number*, Page Range.

ISBN 978-3-0365-4529-5 (Hbk)
ISBN 978-3-0365-4530-1 (PDF)

Contents

About the Editor

Stanisław Duer

Stanisław Duer received his B.Sc. and M.Sc. degrees in electrical engineering from the Military University of Technology, Warsaw, Poland. In 2003, he defended his Ph.D. thesis on technical diagnostics and received a Ph.D. degree from the Department of Mechatronics, Military University of Technology. Since 2003, he has been an Assistant Professor in Applied Electrical Engineering and Electronics in the Department of Mechanics, Technical University of Koszalin, Poland. He is the author and co-author of 16 books and over 190 scientific publications in Neural Computing and Applications, Defence Science Journal, Expert system and Applications, and Energies. Since 2013, he had been working at the Department of Energy in the Faculty of Mechanics at the Technical University of Koszalin as a Professor. His research interests include using big data in power systems, diagnostic systems with an artificial neural network, mathematical modeling, application of mathematical, expert systems, reliability engineering and system safety, innovation in electronic applications in the wind power plants equipment, cars, and others. He is a Guest Editor of the Energies Special Issue 'Intelligent systems supporting the use of energy systems and other complex technical objects, modeling, testing, and analysis of their reliability in the operation process'.

Preface to "Intelligent Systems Supporting the Use of Energy Systems and Other Complex Technical Objects, Modeling, Testing and Analysis of Their Reliability in the Operation Process"

The book focuses on research on intelligent systems supporting the operation of power systems and other equally complex technical facilities. In the technological development of complex technical objects such as aircraft, energy systems, medical devices and others, intelligent assistance systems occupy a special place. Currently, intelligent consulting systems are used to supervise an effective use of technical objects and to organize their maintenance processes. Intelligent systems are particularly useful in technical object diagnosis realized by the qualitative assessment of their reliability. That is especially important for the analysis of highly complex maintenance processes being dependent upon many variables and impacted by non-linear or stochastic factors/conditions. In such a case, testing the reliability condition of a technical device as well as selecting tools and methods for renewing the object (restoration of its operational functions) can be greatly supported by intelligent systems. The necessity of securing precise and well time-bound maintenance (serviceability) is crucial for technical objects operating in a continuous mode (energy generation/processing, security systems) or realizing life-sustaining functions (medical devices, aerospace or automotive/transportation systems).

Special thanks to all the authors involved, and much appreciation for the help of others who supported the work of my scientific teams.

Stanisław Duer
Editor

Article

Determining Information Quality in ICT Systems

Marek Stawowy [1,*], Stanisław Duer [2], Jacek Paś [3] and Wojciech Wawrzyński [1]

[1] Faculty of Transport, Warsaw University of Technology, 75 Koszykowa St., 00-662 Warsaw, Poland; wojciech.wawrzynski@pw.edu.pl
[2] Department of Energy, Faculty of Mechanical Engineering, Technical University of Koszalin, 15-17 Raclawicka St., 75-620 Koszalin, Poland; stanislaw.duer@tu.koszalin.pl
[3] Faculty of Electronic, Military University of Technology of Warsaw, 2 Urbanowicza St., 00-908 Warsaw, Poland; jacek.pas@wat.edu.pl
* Correspondence: marek.stawowy@pw.edu.pl

Abstract: The article deals with the estimation of information quality (IQ) in information and communication technologies (ICT) systems. A number of recent publications were analyzed, as well as ISO standards concerning quality and information quality. Due to the limitations of the known methods of estimating IQ, the authors present their own proprietary concept based on multidimensional and multi-layer modeling using methods of estimating uncertainty. The modeling proposed in this publication uses sixteen dimensions of quality known from the literature. The features of dimensions are taken into account as another layer and information states as successive steps in the IQ model. An example of calculations is also presented in which the mathematical evidence method used in estimating the uncertainty is extended to the modeling of dependent elements. The article also presents a simulation based on the presented example. This simulation shows the assumed dependencies between the output and input values.

Keywords: information quality (IQ); modeling; uncertainty; information and communication technologies (ICT)

Citation: Stawowy, M.; Duer, S.; Paś, J.; Wawrzyński, W. Determining Information Quality in ICT Systems. *Energies* **2021**, *14*, 5549. https://doi.org/10.3390/en14175549

Academic Editor: Silvio Simani

Received: 10 July 2021
Accepted: 1 September 2021
Published: 5 September 2021

Publisher's Note: MDPI stays neutral with regard to jurisdictional claims in published maps and institutional affiliations.

1. Introduction

Nowadays, assessing information quality (IQ) seems to be the first step in assessing a technical system. This is especially the case when it comes to ICT systems in areas such as transport, healthcare, or other emergency services, where the safety of people or transported people and loads frequently depends on the quality of the information provided. Assessing IQ seems to be the first step in assessing the energy supply of ICT systems too. IQ has been studied by many institutions worldwide. One of them launched the Massachusetts Institute of Technology Information Quality (MITIQ) program—an IQ research program carried out after 2000 at MIT [1]. In publications such as [2], the components of IQ dimensions called features are mentioned many times, but hardly any indication is made regarding how to determine these dimensions and features. Additionally, in the ISO 8000 standards [3] there is no reference to the IQ stored dimensions and how to define them. The authors of this article propose a method for determining IQ based on a multidimensional model based on uncertainty modeling. The model takes into account not only the features and dimensions of quality but also information states, due to which the model enables a comprehensive description of the entire ICT system.

2. State of the Art

According to the definition given in [3,4], information is knowledge about objects, such as facts, events, things, processes, or ideas, including concepts that have special meaning in a specific context, and knowledge that reduces or removes the uncertainty about the occurrence of a specific event from a given set of possible events. In turn, data stand for a reinterpretable representation of information that is formalized in a manner suitable for communication, interpretation, or processing. This suggests that data are items of information without a dimension attribute. For example, the detector transmits data about a measured physical quantity with a specific value, but only the attribute of the measurement unit transforms the data into information.

IQ can be called a coefficient (or set of coefficients) which indicates the value of IQ. The definition in the ISO standard includes three components [3]:

1. Syntactic IQ is the degree to which the data conform to a specific syntax;
2. Semantic is the unique and unambiguous conformance between identifiable data units and the entities represented;
3. Pragmatic quality is conformance with the requirements concerning the use.

In this study, the applied methods enable the determination of IQ both in continuous modeling, presented as a set of IQ dimensions [2], and in discrete or hierarchical modeling, but also when divided into categories as in ISO 8000-8 [3]. The applied methods are based on the calculation of dependent and independent coefficients. This approach enables any modeling of multi-layer models (including hierarchical models adopted in the aforementioned ISO standard). Therefore, the rest of this article presents flat IQ models without categorizing. Such a division does not affect the modeling results of independent elements. Because these flat models can be combined into larger multi-layer structures which reflect hierarchical division as well, it can be said that the model described in ISO is one of the particular forms of multi-layer models.

Based on the work of the ancient philosophers Lao Tsu and Plato [5], the measure of quality can be expressed as the pursuit of perfection. Figure 1 shows quality improvement (a term defined in ISO 9000: 2015 [6]) as the pursuit of excellence. The graph in Figure 1 provides the important information that excellence is not achievable. It is the limit of the infinite quality improvement function. However, subsequent steps in improving the quality result in quality improvement (a measure of quality) and the approach to perfection.

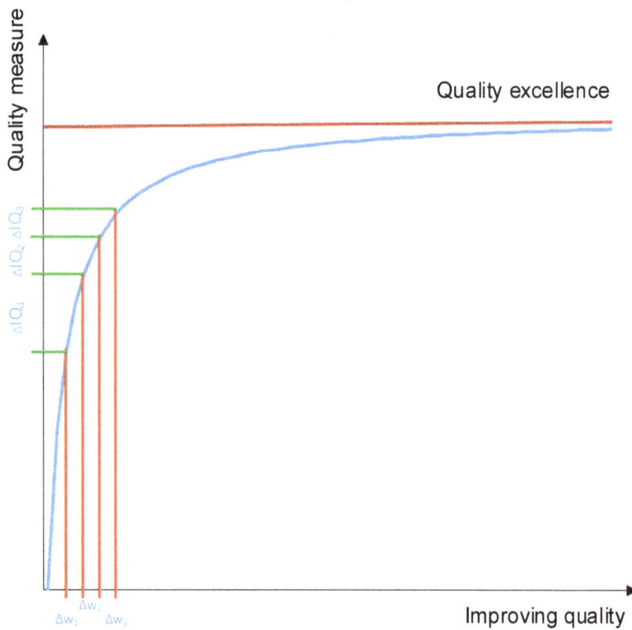

Figure 1. Improving quality as a pursuit of perfection, $\Delta w_1 = \Delta w_2 = \Delta w_3$ and $\Delta IQ_1 > \Delta IQ_2 > \Delta IQ_3$ (own elaboration based on [7]).

Expressed in mathematical symbols:

$$\lim_{n\to\infty} f(w_n) = D \tag{1}$$

where:

D—perfection;

n—quality improvement steps;

w_n—quality measure value string.

Based on Equation (1), it can be assumed that any function converging to some value at infinity can be a function that describes the quality measure well. The same dependencies also apply to IQ.

3. IQ Measure

In technical systems, IQ [3] has a large impact on the assessment of the system, especially when it comes to information systems used in such critical areas as healthcare, energy, and transport. All data can be affected by various errors that introduce a certain amount of uncertainty into the information. This uncertainty may indicate to us the IQ. Hence, the smaller the uncertainty, the higher the IQ. This may concern road traffic control or the uninterruptible power supply (UPS) of hospital equipment. An example of this is the communicativeness and legibility of signs controlling road traffic.

In the literature, there are many studies examining IQ. The following are those that seem to be specific and also characteristic of the subject matter of this article. The main generator of the publication was a project launched at the Massachusetts Institute of Technology under the name MIT Information Quality Program (MITIQ) [1]. Publications related to this project are the most frequently cited references in contemporary IQ studies. There are also many references to publications from this project in this paper. Two books are some of the most important items published under the MITIQ project. The first is "Information Quality" [8]. The book describes the multidimensionality of IQ, how to measure it, and how to manage it. Difficulties in scaling and interpreting IQ measurements are also described. This book mentions as many as one hundred and eighteen quality attributes (features) that can be included in fifteen dimensions of IQ. The second book published within the MITIQ program worth mentioning in this study is "Introduction to Information Quality" [2]. The book discusses the basics related to the currently understood multidimensionality of IQ, which is the basis for the proposed modeling methods in subsequent sections of this paper. The above-mentioned book also includes an extension of the previously crystallized view on the multidimensionality of IQ. It was supplemented, among others, with accessibility, security, and ease of manipulation. The book is the foundation for the guidelines published in 2010 on the US Department of Justice's website regarding IQ.

In 2015, the ISO 8000-8 Information and Data Quality: Concepts and Measuring standard [3] was published partly as an alternative to the work of MITIQ. The standard rather modestly refers to what was developed by the MITIQ program, but cites authors from earlier publications, including the authors of the two above-mentioned books [2,8]. ISO 8000-8 is a set of standards describing the IQ dimensionality and hierarchical classification of IQ, and includes definitions of basic concepts such as: data, information, metadata, and data unit. This standard introduces a division of IQ into three main categories: synthetic, semantic, and pragmatic. It also schematically presents the general principles of measuring the overall IQ and the specific (subjective) IQ for each category. The model presented there is quite modest and its description does not reflect the real extent of the problem but only tries to standardize the approach, which by using the above-mentioned three categories seems quite limited. In the following sections, the concept of determining IQ is developed based on a multi-layer quality model. The model described in ISO 8000 is also compatible with the concept described; namely, it is its special case.

Successive publications appearing worldwide indicate the extent and variety of issues related to IQ, its determination, and interpretation. Publication [9] is a fine example. This publication presents approaches to IQ analysis in data integration framework schemes. It describes the integrated scheme quality assessment and distributed access to information systems, focusing on the minimum, consistency, and completeness of information. A multidimensional model of IQ was used for the evaluation. This approach is quite similar to that proposed in the following sections, although quite modest.

The publication [10] introduces a new approach to IQ measurement and uses the Six Sigma (6σ) method to estimate IQ. This approach focuses on continuously improving the IQ by systematically assessing many of the IQ dimensions. In particular, it deals with the correlation and the relative importance of the IQ dimensions. Thanks to this method, a precise and systematic criterion for assessing the quality of information is proposed. This concept seems quite attractive, but it does not exhaust the complexities of IQ modeling.

The article [11] discusses and analyzes the notion of IQ in terms of a pragmatic philosophy of language. It states that the concept of IQ is of great importance and must be situated better within a sound philosophy of information. It turns out that much research on IQ conceptualizes IQ as an inherent property of the information itself. A model of multidimensional IQ was presented, in which twenty-two dimensions were specified (accurate, appropriate, authentic, authoritative, balanced, believable, complete, comprehensive, correct, credible, current, good, neutral, relevant, reliable, objective, true, trustworthy, understandable, useful, usability, valid). These are more than the dimensions used in modeling conducted in the following sections. However, the modeling proposed in the following sections is open-ended and can theoretically be applied to an indefinite number of dimensions and may also include the number of dimensions shown in the article [11].

One of the co-authors (the main co-author) published the first original publications describing the IQ in 2013 and 2014 [12,13]. The articles present a model for determining IQ based on the Certainty Factor (CF) in highway telematics. Such modeling usually concerns expert systems or artificial intelligence. However, in highway telematics systems, the main elements are computer systems that analyze and process data on vehicle traffic. Modeling assumed multidimensionality of the IQ. These dimensions are shown as both dependent and independent.

In [14], the authors present the estimation of IQ in various domains. The article discusses the issues of portability of IQ modeling between domains. This led to the conclusion that an independent model should actually be created for each domain. The arguments presented in the following sections of this study lead to similar conclusions. In the proposed method of multidimensional, open modeling in this article, it is possible to build such an open model that will enable the description of IQ in many domains.

In 2014, the main co-author published two original works that provide the basis for this study [15,16]. Both publications were presented at the ESREL (European Safety and Reliability) Conference in 2014. The first paper [15] discusses the IQ estimation model of ICT systems based on CF modeling. This modeling practice was typically used in expert systems or artificial intelligence. Here, however, computer systems that use data from ICT systems are discussed. CF modeling is one of the methods that allow us to obtain information about the properties of a system when data about this system are incomplete. The model helps identify and locate the weakest system components that have a disastrous effect on IQ. The second publication [16] is a continuation of the previous works [12,13] involving the determination of IQ in various systems. When describing IQ, several basic dimensions were defined, such as: availability, actual value, completeness, reliability, flexibility, form, importance over time, accuracy, reliability, selectivity, and importance. One of the features of IQ dimensions was determined. The CF-modeling and Dempster–Shafer mathematical evidence methods were used.

The discussion on this topic was extended by the co-authors at the next ESREL 2015 conference [17]. The publication demonstrated that modeling the uncertainty of IQ can be achieved using the mathematical evidence theory as in the publication from ESREL 2014 [15,16]. While in the case of independent sources influencing the IQ, the use of evidence theory is quite simple, in the case of dependent sources, this modeling is not possible. This work proposes a method of determining the IQ for dependent sources (a serial model). A two-layer model consisting of dependent and independent elements was presented. This multi-layer modeling became the basis for the models shown in the following sections of this study.

A different approach can be seen in the study described in [18]. It attempts to investigate the importance of many information dimensions in measuring the IQ from the user's point of view. The article provides a detailed analysis of the nature and importance of the various dimensions of IQ and their differences depending on the context and user demographics. This is an approach that takes into account only the subjective dimensions of IQ.

The next work [19] presents the issues of IQ measurability. The article examines the reasons underlying the differences in the measurability of IQ. Using the structure of Gigerenzer's "building blocks", it was hypothesized that the feasibility of using a set of heuristic principles when assessing different IQ dimensions is a key factor influencing the inter-rater agreement (content moderators) in IQ judgments. This method was used to assess IQ in Internet resources.

Alternative approaches to understanding and modeling IQ that typically involve a particular approach or partial quality assessment have been described above. The publications below display how IQ can be measured. This issue has been examined not only for studies related to technical systems.

In [20], the methodology for assessing the IQ for fifteen dimensions was defined and arranged in groups. The proposed methodology for IQ assessment (AIMQ) as a whole provides a practical tool for measuring IQ for an organization. It can apply at various organizational levels, such as the financial industry, healthcare, and manufacturing. The methodology is useful in identifying IQ issues, prioritizing areas of IQ improvement, and monitoring IQ improvements over time. This article presents a method that allows the IQ to be assessed in a hierarchical practical model arranged in groups. Such modeling usually has limitations; for example, such a model cannot be open because it is limited by groups. It has the same restrictions as the model described in ISO 8000 [3].

The article [21] presents a method that can be used in measuring the IQ of Internet resources. The presented method of measuring the IQ was limited to sixteen criteria, which partially overlapped with the dimensions presented in [2]. The method was based on four successive steps with repetitions of sections. The content of websites, traffic volume, understanding, and feedback were examined, which means that this method enables the measurement of the quality of both information content and the quality of the medium that the Internet is.

A different approach was presented in [22]. It attempts to indicate the best method of quality measurement yet, assuming that it is the definition of quality that imposes the measurement methodology. The paper includes a literature review and detects flaws in the methods presented there in the form of omitting the variability of requirements over time and different meanings of quality features. A method was proposed based on the division into analytical and synthetic measurements.

The study in [23] proposed the quality assessment on two levels. A quality assessment based on an information decomposition of the fusion system in its elementary modules was planned. The first (global), which describes the entire information fusion system, and the second (local), for each elementary module. The method was based on the multidimensionality of the IQ, and the fusion was performed by estimating the Bayes' subjective probability. The method seems very complicated, which limits its use.

The following article, [24], presented an IQ model that shows how to understand IQ in the context of systems and also how to determine some common IQ indicators. The importance of predicting and modeling the IQ was also described. Building information chains automatically to meet the expected IQ was suggested. The limitation of this method is the application of chains that prevent the use of more complex structures.

Uncertainty modeling to determine IQ occurs very rarely in the literature. One of the few examples which do not belong to the authors of this paper is [25] and describes the relationship between IQ and uncertainty modeling. Information uncertainty was presented as part of the IQ model. This type of approach seems to be very attractive, as shown in the following sections. However, the article does not develop a method into full IQ modeling using uncertainty modeling.

Summarizing the above-mentioned publications, it seems obvious that there is no specific method for determining the IQ, especially in technical systems. The authors of this article are trying to fill in the lack of such a study. Another disadvantage of the methods proposed above is the fact that they are often limited to a specific model. That is, the lack of openness in modeling to new dimensions of quality or their features. Another restriction of the presented methods is often the dependence on individual groups of IQ dimensions, which limits the flexibility of modeling and confines this modeling to selected dimensions. Another limitation is the frequent omission of the possibility of multi-level modeling with even the simplest, minimal division of quality dimensions into features. There is also no link between the IQ and its subsequent states. The authors make an attempt to eliminate all these limitations in this study, presenting a multi-layer model of IQ using uncertainty modeling and taking into account information states known from the literature. The article uses the mathematical evidence method as an example of uncertainty modeling used to determine the IQ. In the following sections, in addition to the description of the method, an example of calculations for the selected model and the simulation of the model results depending on the input coefficients are also presented. A similar approach was presented by the main co-author in [7,26–29]. Applying the presented approach to the assessment and analysis of other systems by assessing reliability or risk, as presented in the works [30–32], is also expected. To use the modeling presented here in various other types of technical system assessments, such as diagnostics [33], risk assessment related to road and rail signaling [27,34], and development-related applications [35], seems possible too.

4. Research Problem and Research Methodology

On the one hand, in the definition of the problem, the information quality dimensions can be distinguished, which depend on their features. On the other hand, there might appear information states, which depend on the structure of the ICT system. In order to model information quality, a flat model of sixteen quality dimensions presented in [2] has been adopted [7,8,27]. This model exhibits great elasticity and enables us to define the features of quality dimensions and also to subordinate the dimensions from these features. This model applies the quality dimensions, which are presented in Table 1 and in Figure 2.

Table 1. Quality dimensions. Based on [7].

No.	Name of the Dimension	Dimension's Interpretation
1	Availability (D_{av})	This dimension determines the possibility of exploiting the ICT element on demand in a given time and by using an authorized process. This dimension is directly related to the security of information.
2	Appropriate amount of data (D_{aad})	This dimension determines what amount of data are appropriate to enable task execution and simultaneously indicates that the given amount is sufficient and that more data could decrease the quality of information.
3	Believability (D_{bel})	This dimension determines the degree to which information reflects reality. It can also be related to the believability of the source of information.
4	Completeness (D_{com})	This dimension determines whether the data are sufficient to execute a particular task.
5	Concise representation (D_{ccr})	This dimension determines the degree to which data are represented.
6	Consistent representation (D_{csr})	This dimension determines the degree to which data are represented with the same size.
7	Ease of manipulation (D_{eom})	This dimension determines how easy it is to process these data for different task applications.
8	Free of error (D_{foe})	This dimension determines to what degree data are free of error.
9	Interpretability (D_{inter})	This dimension determines the degree to which data are clear and represented in appropriate languages and symbols.
10	Objectivity (D_{obj})	This dimension determines to what degree data are not subjective, i.e., limited to a narrow scope.
11	Relevancy (D_{relev})	This dimension determines the degree to which the data are applicable to this particular task.
12	Reputation (D_{reput})	This dimension determines the degree to which data are evaluated for their source and content.
13	Security (D_{sec})	This dimension determines the limitation of data access in order to ensure security and protect from unauthorized access.
14	Timeliness (D_{tim})	This dimension determines the degree to which data are available on time in order to execute the task.
15	Understandability (D_{uns})	This dimension determines the data's understandability.
16	Value-added (D_{vadd})	This dimension determines the advantages of exploiting data and whether the data are beneficial for task execution.

The second element which demands modeling is the structure of the ICT system. One can encounter in the literature many models describing diverse information states in systems. The ones mostly elaborated on occur in [36], where they are called information processes. The following types of information processes can be specified: generating, collecting, storage, processing, transmission, sharing, and interpreting (Figure 3).

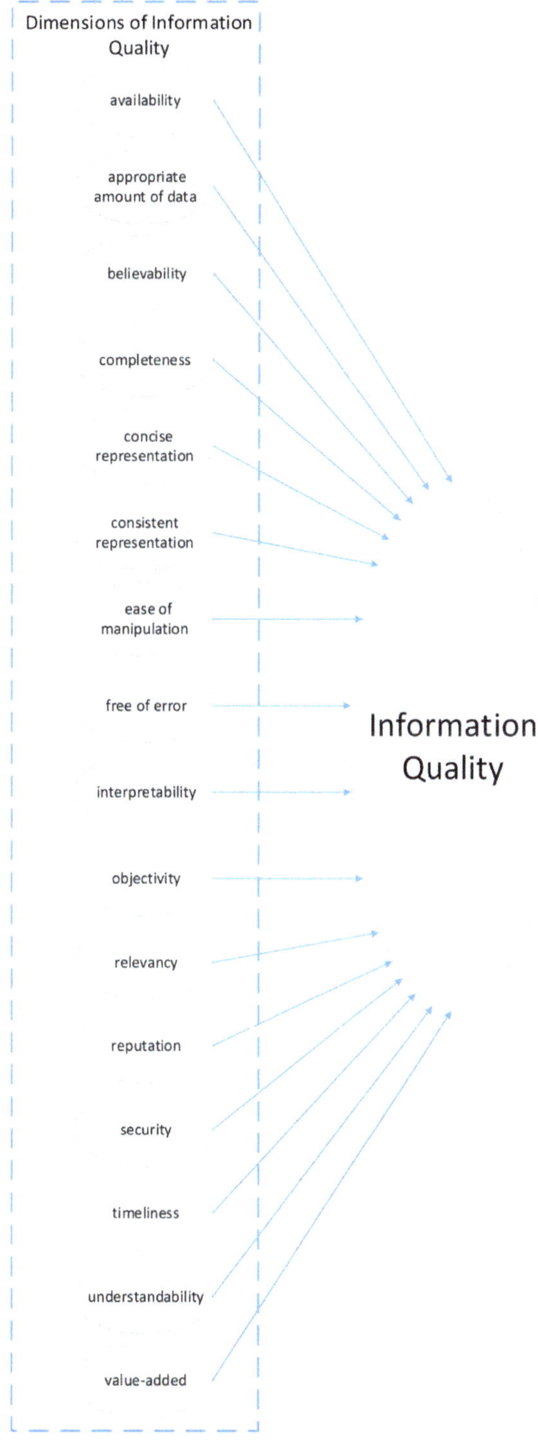

Figure 2. Information quality dimensions. Own elaboration based on [7].

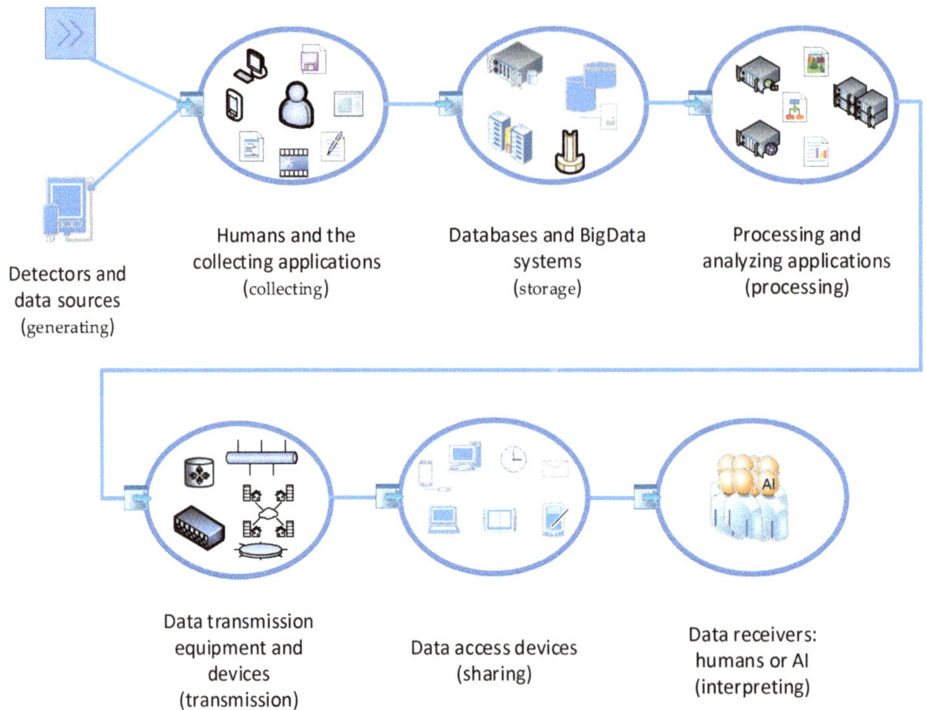

Figure 3. States related to registration and data transmission in an ICT system. Own elaboration based on [7,27,29].

Each of the seven above-named information states is a consecutive element influencing IQ. Thus, a formula can be devised in the following way:

$$IQ = f(w_{11}, w_{12}, \ldots, w_{1m}, w_{21}, w_{22}, \ldots, w_{2m}, \ldots w_{nm}) \tag{2}$$

where:

m—the number of dimensions, IQ components (equals 16 according to Table 1);

n—the number related to registration and data transmission in an ICT system (equals 7 according to Figure 3);

w—a variable determining the influence of the particular dimension (e.g., range of values [0,1]).

The general form of the matrix:

$$IQ = \begin{bmatrix} w_{11} & \cdots & w_{1m} \\ \vdots & \ddots & \vdots \\ w_{n1} & \cdots & w_{nm} \end{bmatrix} \tag{3}$$

As it has been mentioned before, searching for a method to determine IQ dimensions appears obvious. Figure 4 presents the positioning of the quality dimensions features in respect to the dimensions themselves. It is worth noting that the dimensions as such can be completely independent, yet their features can be shared between the dimensions. This means that one feature can influence many IQ dimensions. For example, dimensions (see Table 1) no. 3 (believability) and no. 8 (free of error) can have mutual determining features, e.g., errors of transmission or of data storage, which constitute the information (Figure 5).

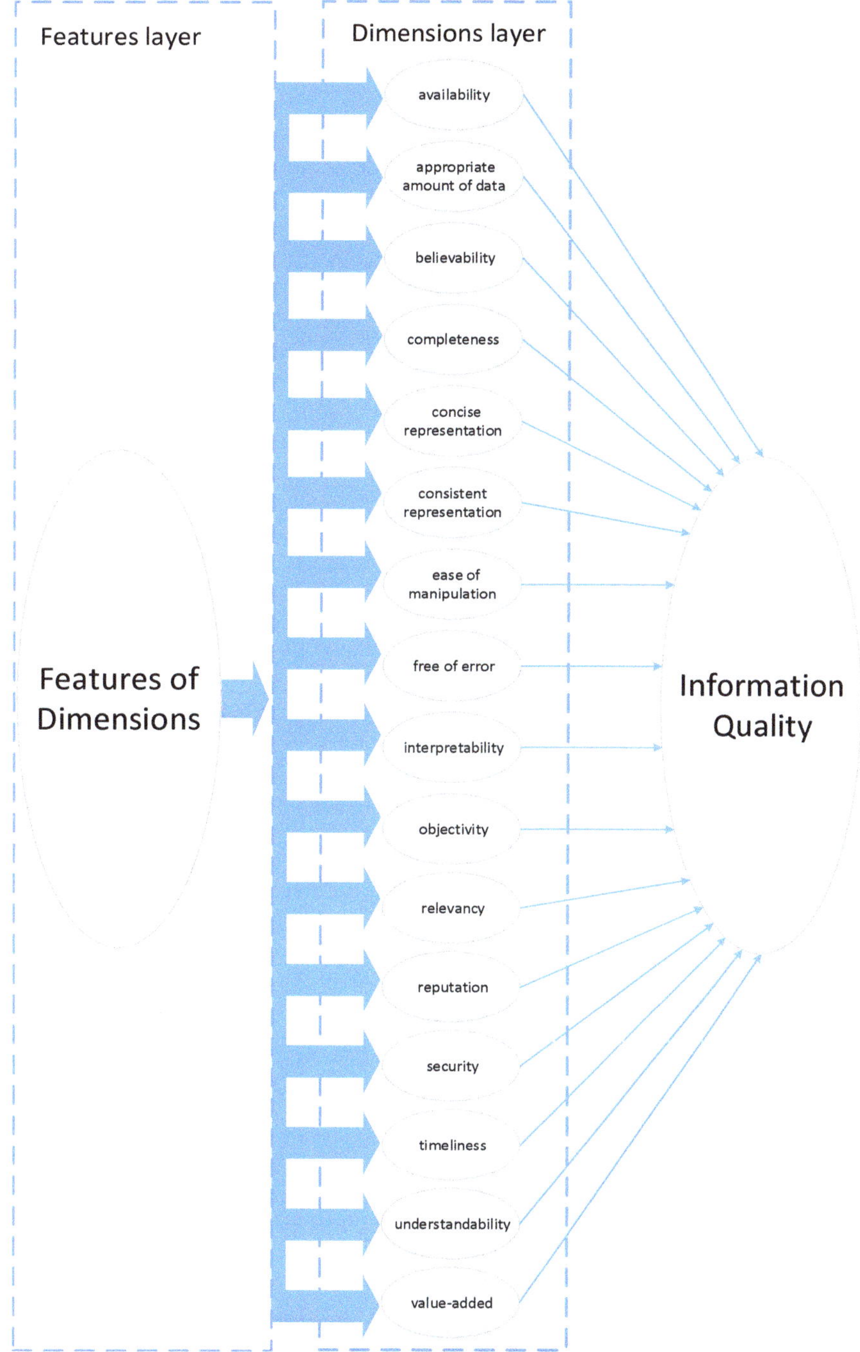

Figure 4. A diagram of IQ model extension with its dimension features. Own elaboration.

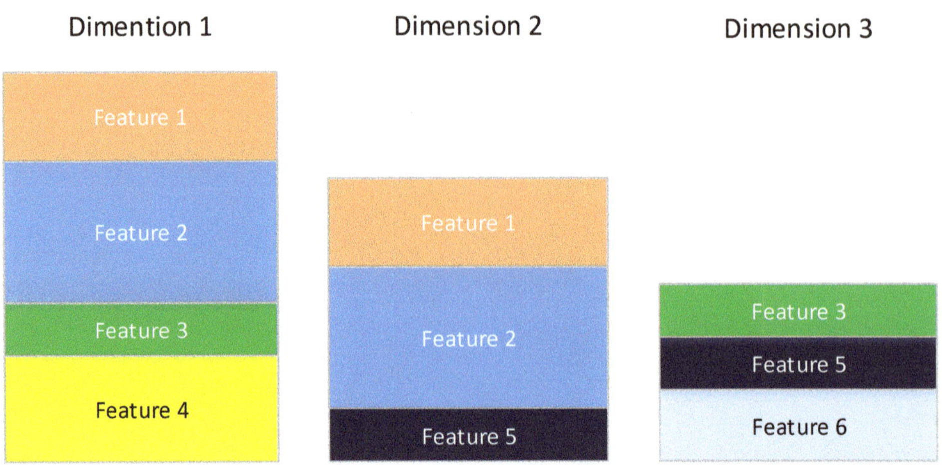

Figure 5. A diagram showing dependencies between dimensions on the features layer. Own elaboration.

Figure 6 presents a universal diagram of an IQ model for an ICT system taking into account information states, IQ dimensions, and their dimension features.

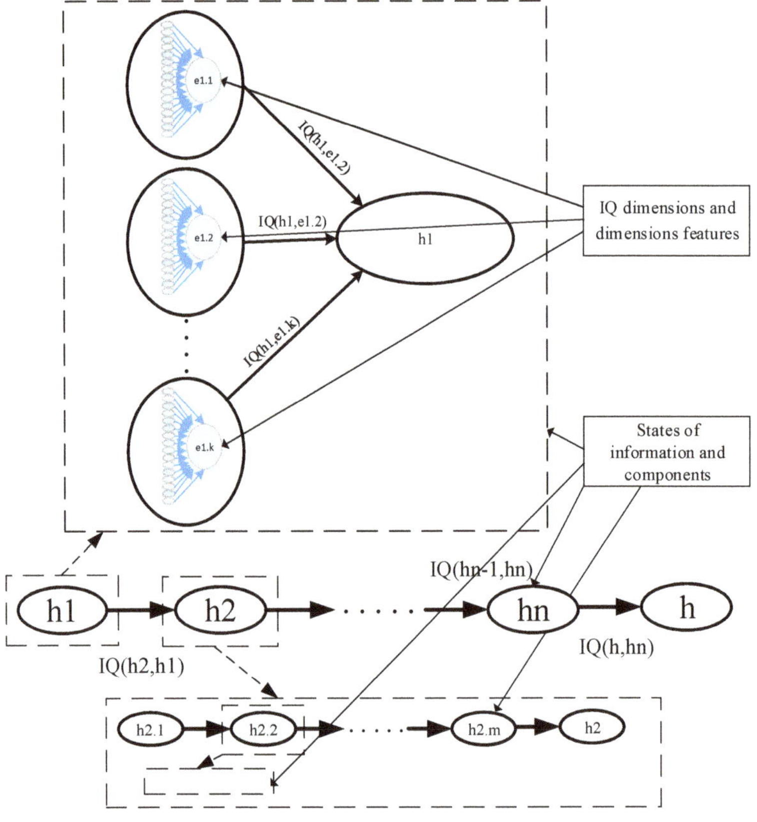

Figure 6. A diagram of IQ model extension. Own elaboration based on [7].

A method to establish the value of factors of quality dimension features served as a method devised to determine IQ dimensions in publication [17]. This method is based on uncertainty modeling using mathematical evidence and dependent relations in serial models.

Taking into consideration what has been so far described in this study, when determining the IQ of a chosen ICT system, one should follow the flowchart in Figure 7. The first step is the choice of stages of information of the given ICT system on the basis of Figure 3 (first step in Figure 7). The second step is the choice of IQ dimensions on the basis of Figure 2 or Table 1. At this point, all dimensions can be taken into account, but this will complicate the calculations. Generally, it is not necessary to include all IQ dimensions presented in Table 1 to determine the IQ of a chosen system. In the third step, the features of the chosen dimensions (Figure 4) should be selected allowing for the fact that one feature can affect several dimensions (Figure 5). According to the literature [2], over one hundred and thirty features can be assigned to the presented sixteen dimensions. Only those features that have a significant impact on IQ, as in the example, should be selected. In the fourth step, it is possible to decide which of the information stages for the evaluated ICT system will be multiplied (Figure 6). The penultimate step is to create a model or models to describe the impact of subsequent elements on the IQ. The last step is the calculation leading to one final IQ indicator. This sequence is used in the example in Section 7.

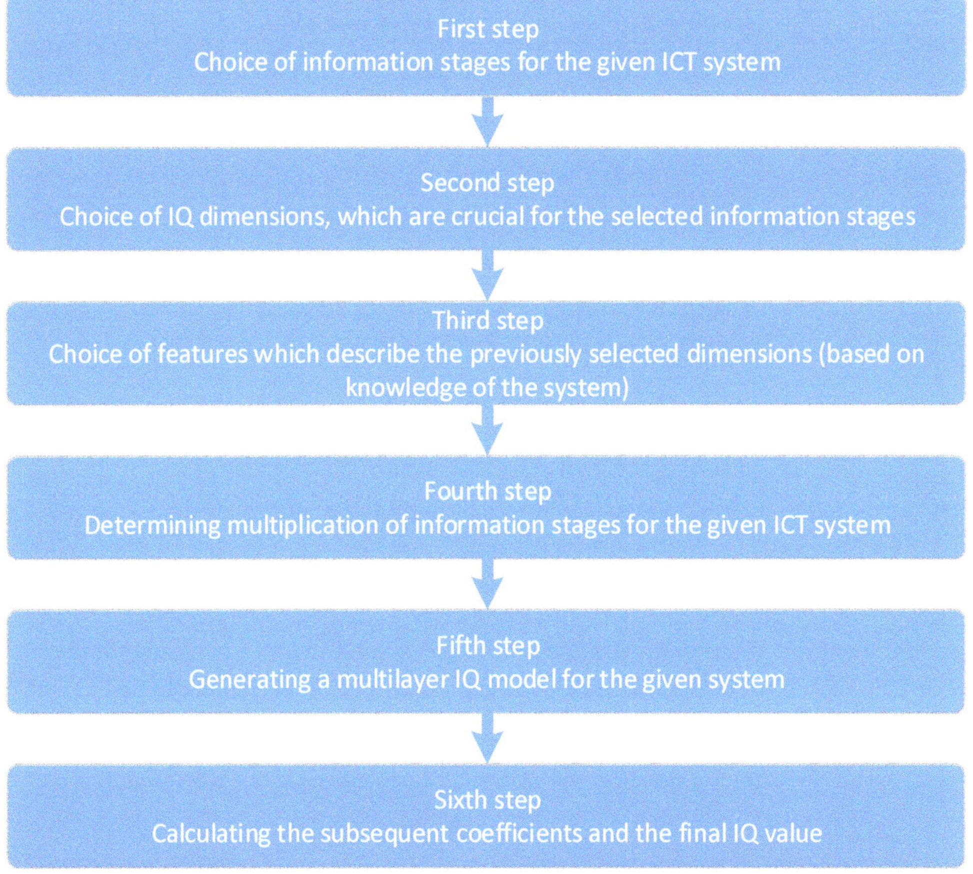

Figure 7. Flowchart on the basis of the described method. Own elaboration.

5. Uncertainty Modeling on the Basis of the Theory of Mathematical Evidence

In the theory of mathematical evidence, it is possible to synthesize information for individual elementary probability measures. They can be synthesized even if they are contradictory or come from different sources [16,17,26,27,37,38]. This enables the synthesis of independent information. Such a synthesis can be defined by the formula:

$$m_3 = \frac{\sum_{A_i \cap B_j = C}(m_1(A_i) \cdot m_2(B_j))}{1 - \sum_{A_i \cap B_j = \varnothing}(m_1(A_i) \cdot m_2(B_j))} \tag{4}$$

where A, B, and C—sources of observation that represent subsets of the set Θ; m_1 and m_2—sets of masses; and m_3—a new set of masses, defined by Dempster as a certain degree of faith or subjective probability [37].

This synthesis is called the Dempster combination rule [27,38]. The Basic Belief Assignment (BBA) function is defined as follows:

$$m : 2^{\Theta} \to \langle 0,1 \rangle$$
$$m[\varnothing] = 0 \tag{5}$$
$$\sum_{A \subseteq \Theta} m(A) = 1$$

Belief, abbreviated as Bel $\in$ [0,1], measures the strength of the obtained observations supporting the belief that the set of hypotheses is true.

$$Bel(A) = \sum_{B \subseteq A} m(B) \tag{6}$$

Plausibility, abbreviated as Pl $\in$ [0,1], determines to what extent the belief about the truth of A is limited by the supporting evidence $\neg A$.

$$Pl(A) = \sum_{B \cap A \neq \varnothing} m(B)$$
$$Pl(A) = 1 - Bel(\neg A) \tag{7}$$

Doubt, in short Dou $\in$ [0,1], measures the strength of the obtained observations supporting the doubt as to the authenticity of the examined set of hypotheses.

$$Dou(A) = 1 - Bel(A) \tag{8}$$

Disbelief, abbreviated as Dis $\in$ [0,1], determines the extent to which the doubt of A's authenticity is limited by the supporting evidence of $\neg A$.

$$Dis(A) = 1 - Pl(A) \tag{9}$$

The combination rule affects the belief function and can be expressed like this:

$$Bel1 \oplus Bel2(A) = \sum_{B \subseteq A} m_1 \oplus m_2\,(A) \tag{10}$$

6. Multi-Layer Modeling of Uncertainty Using the Hybrid Method

Multi-layer modeling demands serial connections (dependents). A mathematical description of dependent sensors which record observations was suggested in [39]. The study resulted in a formula based on a Cartesian product in the form of:

$$m = m_1 \otimes m_2 \otimes \ldots \otimes m_n \tag{11}$$

where $m_1, m_2 \ldots m_n$ sets of masses; m new set of mass.

Yet, in the case of the modeling presented in this article, the hypothesis factors on the dependent layer will be described on a single dimension. Thus, the formula can be simplified as follows:

$$m_3 = m_1 \cdot m_2 \tag{12}$$

7. Method Demonstration

To demonstrate the method, a program was written simulating some features of one of the IQ dimensions, and the target final quality value for the model of the IQ ICT system is presented in Figure 8. This software was created by Marek Stawowy, one of the authors of this article. The software under the name DSHyb enables calculations for uncertainty models applying DS (mathematical evidence) and the hybrid method. From Figure 8, it is evident that the model was restricted to two states of information of an ICT system. State e1 stands for the state of information transmission, and state e2 stands for the state of information interpretation.

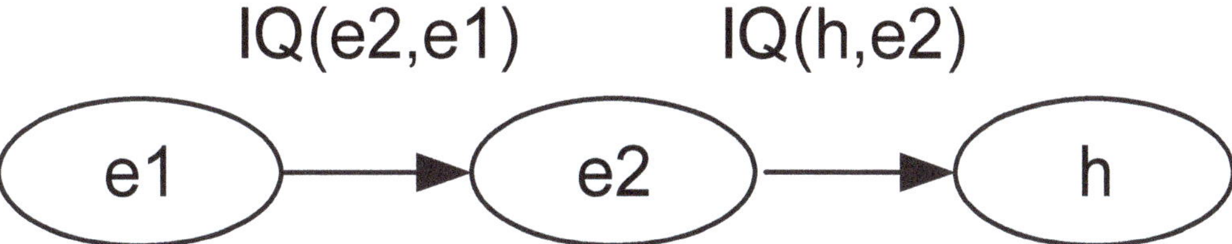

Figure 8. Model of ICT system quality as used in the example. Own elaboration.

The model of IQ state e1 presented in Figure 9 includes three IQ dimensions:

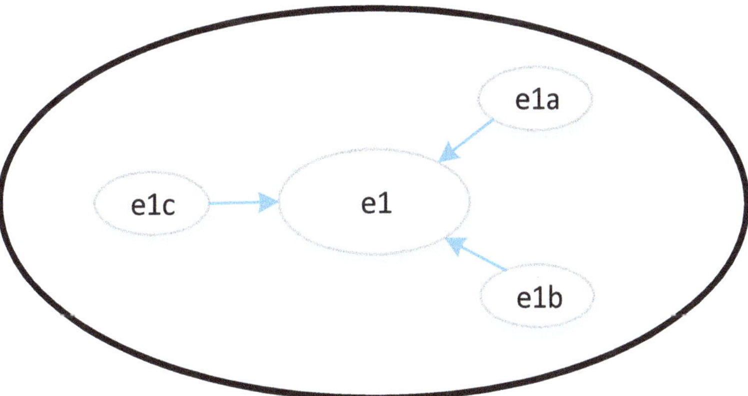

Figure 9. Model of IQ state e1. Own elaboration.

e1a—appropriate amount of data;
e1b—believability;
e1c—error free.
The model of dimension e1c presented in Figure 10 includes four dimensions features:

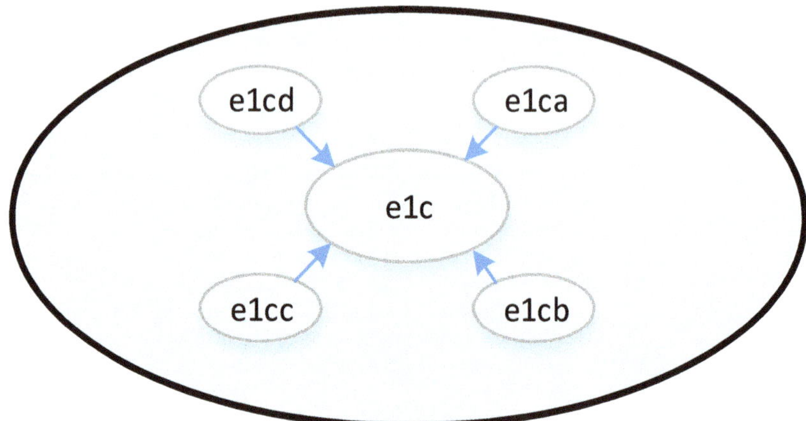

Figure 10. Model of IQ dimension e1c. Own elaboration.

e1ca—correct information transmission;
e1cb—transmission of faulty data;
e1cc—data assignment to attributes failure;
e1cd—wrong attributes.
Table 2 presents values that are assigned to the model of IQ dimension in Figure 9.

Table 2. Values assigned to the quality dimension features for the model presented in Figure 9. Own elaboration.

Observation	Value
e1ca	0.99
e1cb	0.001
e1cc	0.002
e1cd	0.0005

A table is the clearest way of presenting calculations for independent elements using the theory of mathematical evidence. Thus, presented in this form, it is possible to show dependencies in observation tables, as shown in Tables 3–6.

$$\Theta = \{e1ca, e1cb, e1cc, e1cd\} \tag{13}$$

$$m_1\,(\Theta) = 1 \tag{14}$$

Table 3. Observation e1ca.

$m_2\,(\{e1ca\}) = 0.99$ $m_2\,(\Theta) = 0.01$	$m_2\,(\{e1ca\})$	$m_2\,(\Theta)$
$m_1\,(\Theta)$	$m_3\,(\{e1ca\})$	$m_3\,(\Theta)$

Table 4. Observation e1cb.

$m_4\,(\{e1cb\}) = 0.001$ $m_4\,(\Theta) = 0.999$	$m_4\,(\{e1cb\})$	$m_4\,(\Theta)$
$m_3\,(\{e1ca\})$ $m_3\,(\Theta)$	$m_5\,(\{\varnothing\})$ $m_5\,(\{e1cb\})$	$m_5\,(\{e1ca\})$ $m_5\,(\Theta)$

Table 5. Observation e1cc.

$m_6\ (\{e1cc\}) = 0.002$ $m_6\ (\Theta) = 0.998$	$m_6(\{e1cc\})$	$m_6\ (\Theta)$
$m_5\ (\{\varnothing\})$	$m_7\ (\{\varnothing\})$	$m_7\ (\{\varnothing\})$
$m_5\ (\{e1cb\})$	$m_7\ (\{e1cb\})$	$m_7\ (\{e1cb\})$
$m_5\ (\{e1ca\})$	$m_7\ (\{\varnothing\})$	$m_7\ (\{e1ca\})$
$m_5\ (\Theta)$	$m_7\ (\{e1cc\})$	$m_7\ (\Theta)$

Table 6. Observation e1cd.

$m_8\ (\{e1cd\}) = 0.0005$ $m_8\ (\Theta) = 0.9995$	$m_8(\{e1cd\})$	$m_8\ (\Theta)$
$m_7\ (\{\varnothing\})$	$m_9\ (\{\varnothing\})$	$m_9\ (\{\varnothing\})$
$m_7\ (\{e1cc\})$	$m_9\ (\{e1cc\})$	$m_9\ (\{e1cc\})$
$m_7\ (\{e1cb\})$	$m_9\ (\{e1cb\})$	$m_9\ (\{e1cb\})$
$m_7\ (\{e1ca\})$	$m_9\ (\{\varnothing\})$	$m_9\ (\{e1ca\})$
$m_7\ (\Theta)$	$m_9\ (\{e1cd\})$	$m_9\ (\Theta)$

Tables 3–6 show the subsequent stages of mass calculation. As a result, mass m_9 with the index e1ca indicates the Bel result (e1ca).

$$\text{Bel (e1ca)} = m_9\ (\text{e1ca}) \tag{15}$$

The determination of the Bel value can also be presented in the form of a matrix. A detailed example of such an operation can be found among others in the following publications: [17,27].

Having assigned the value with the use of the hybrid method described above and in [17,27], e1c = Bel (e1ca) = 0.9801.

Table 7 presents values that are assigned to the model of IQ dimension in Figure 8.

Table 7. Values assigned to the quality dimensions for the model presented in Figure 8. Own elaboration.

Observation	Value
e1a	0.88
e1b	0.91
e1c	0.9801

Having assigned the value with the use of the hybrid method described in [17,27] and the example for calculating e1c, e1 = 0.99944.

Table 8 presents values that are assigned to the model of IQ states in Figure 7.

Table 8. Values assigned to the information states for the model presented in Figure 7. Own elaboration.

Observation	Value
e1	0.99944
e2	0.98

In this case, the elements are dependent (serial), so Equation (12) must be applied. Thus, h = e1 · e2 = 0.97945.

This value will be the IQ coefficient of the ICT system presented in the above example.

As in the articles [17,29], below is presented a simulation of IQ depending on the positive observation e1ca (Figure 11) and negative e1cb (Figure 12). In order to obtain a visualization of IQ dependency change as a function of one of the IQ dimension features, a simulation was performed for the e1ca (observation with a positive influence on the IQ) value with a range between 0.05 and 0.99. The results of this simulation are presented in the form of a diagram in Figure 11. As a result of the simulation, an approximate relationship was obtained, which is presented as an expected graph in Figure 1.

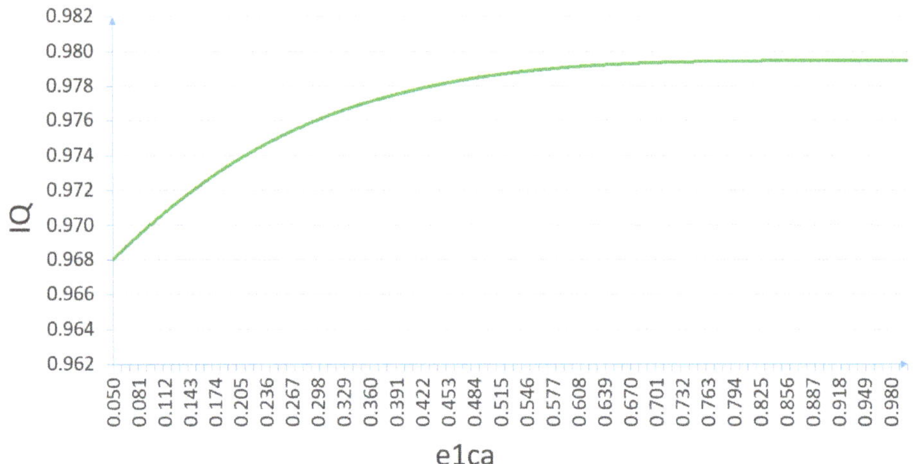

Figure 11. IQ dependency graph as a function of one of the IQ features, e1ca. Own elaboration.

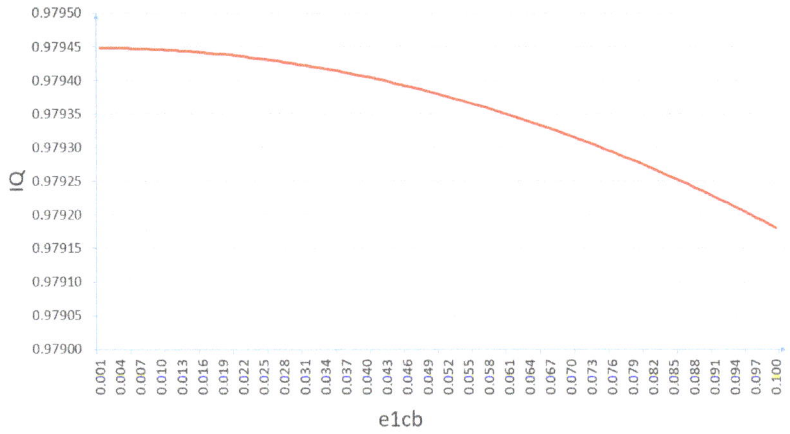

Figure 12. IQ dependency graph as a function of one of the IQ features, e1cb. Own elaboration.

In order to obtain a visualization of IQ dependency change as a function of one of the IQ dimension features, a simulation was performed for the e1cb (observation with negative influence on the IQ) value with a range between 0.001 and 0c. The results of this simulation are presented in the form of a diagram in Figure 12.

8. Conclusions and Summary

This article proposes a method for determining the quality of information (IQ) founded on a multidimensional model based on uncertainty modeling. The model takes into account not only the features and dimensions of quality but also information states, thanks to which the model enables a comprehensive description of the entire ICT system. Therefore, the presented method allows determining multi-layer dependencies both dependent and independent (serial and parallel) in complex IQ models of ICT systems consisting of information states, IQ dimensions, and features of these dimensions. Due to this method, it is possible to narrow down the IQ of the ICT system to one indicator, which can also serve as an evaluation of the system. This article also describes an example and a simulation of the influence of the IQ dimensions on the ICT system IQ. The example uses mathematical evidence as a method to estimate the uncertainty of independent elements extended by the calculation of dependent elements. As a result of the simulation, an approximate relationship is obtained, which is presented as expected (Figures 1 and 11). The next stage in the development of this method might be the application of other ways of modeling uncertainty, because the theory of mathematical evidence is of little practical use in multi-element models. This results from the fact that complications in calculations grow rapidly with regard to the number of independent objects in the model. A different direction for the development of the method presented in this study is devising new methods of calibration depending on the requirements of the given system.

Author Contributions: Conceptualization, M.S., S.D. and J.P.; methodology, M.S. and W.W.; software, M.S.; validation, J.P. and W.W.; formal analysis, S.D. and J.P.; investigation, M.S. and J.P.; resources, M.S. and W.W.; data curation, M.S. and W.W.; writing—original draft preparation, M.S., S.D., J.P. and W.W.; writing—review and editing, M.S. and J.P.; visualization, M.S. and W.W.; supervision, M.S. All authors have read and agreed to the published version of the manuscript.

Funding: This research received no external funding.

Institutional Review Board Statement: Not applicable.

Informed Consent Statement: Not applicable.

Data Availability Statement: Not applicable.

Conflicts of Interest: The authors declare no conflict of interest.

References

1. Massachusetts Institute of Technology Information Quality (MITIQ) Program. Available online: http://mitiq.mit.edu (accessed on 3 June 2021).
2. Fisher, C.; Lauria, E.; Chengalur-Smith, S.; Wang, R. *Introduction to Information Quality*; Authorhouse: Bloomington, IN, USA, 2011.
3. International Organization for Standardization. *Data Quality—Part 8: Information and Data Quality: Concepts and Measuring*; ISO/IEC 8000-8:2015; ISO: Geneva, Switzerland, 2015.
4. International Organization for Standardization. *Information Technology—Vocabulary—Part 28: Artificial Intelligence—Basic Concepts and Expert Systems*; ISO/IEC 2382:2015: 2121272, 2121271 & 2123204; ISO: Geneva, Switzerland, 2015.
5. Tatarkiewicz, W. *History of Philosophy*, 22nd ed.; PWN: Warsaw, Poland, 2014; Volumes 1–3.
6. International Organization for Standardization. *Quality Management Systems—Fundamentals and Vocabulary*; ISO/IEC 9000:2015; ISO: Geneva, Switzerland, 2015.
7. Stawowy, M.; Olchowik, W.; Rosiński, A.; Dąbrowski, T. The Analysis and Modelling of the Quality of Information Acquired from Weather Station Sensors. *Remote Sens.* **2021**, *13*, 693. [CrossRef]
8. Wang, R.Y.; Pierce, E.M.; Madnick, S.; Fisher, C.W. (Eds.) *Information Quality. Advances in Management Information Systems*; M.E. Sharpe: Armonk, NY, USA, 2005.
9. Batista, M.D.; Salgado, A.C. Information Quality Measurement in Data Integration Schemas. In Proceedings of the Fifth International Workshop on Quality in Databases, QDB 2007, at the VLDB 2007 Conference, Vienna, Austria, 23 September 2007.
10. Lee, S.; Haider, A. *A Framework for Information Quality Assessment Using Six Sigma Approach. Communications of the IBIMA*; IBIMA: King of Prussia, PA, USA, 2011. [CrossRef]
11. Mai, J.E. The Quality and Qualities of Information. *JASIST* **2013**, *64*, 675–688. [CrossRef]
12. Stawowy, M. Quality of Information Fed by Video Surveillance Systems Protecting Critical Infrastructure. *WUT J. Transp. Eng.* **2014**, *104*, 103–113.
13. Stawowy, M. Determination of Information Quality of Motorway Telematics. *WUT J. Transp. Eng.* **2013**, *92*, 221–229.
14. Illari, P.; Floridi, L. Information Quality, Data and Philosophy. In *The Philosophy of Information Quality. Synthese Library (Studies in Epistemology, Logic, Methodology, and Philosophy of Science)*; Springer: Cham, Switzerland, 2014; Volume 358. [CrossRef]
15. Stawowy, M. Comparison of uncertainty models of impact of teleinformation devices reliability on information quality. In *Safety and Reliability: Methodology and Applications—Proceedings of the European Safety and Reliability Conference ESREL 2014*; Nowakowski, T., Młyńczak, M., Jodejko-Pietruczuk, A., Werbińska–Wojciechowska, S., Eds.; CRC Press/Balkema: London, UK, 2015; pp. 2329–2333.

16. Stawowy, M. Model for information quality determination of teleinformation systems of transport. In *Safety and Reliability: Methodology and Applications—Proceedings of the European Safety and Reliability Conference ESREL 2014*; Nowakowski, T., Młyńczak, M., Jodejko-Pietruczuk, A., Werbińska–Wojciechowska, S., Eds.; CRC Press/Balkema: London, UK, 2015; pp. 1909–1914.
17. Stawowy, M.; Dziula, P. Comparison of Uncertainty Multilayer Models of Impact of Teleinformation Devices Reliability on Information Quality. In *Safety and Reliability of Complex. Engineered Systems—Proceedings of the European Safety and Reliability Conference ESREL 2015*; Podofillini, L., Sudret, B., Stojadinovic, B., Zio, E., Kröger, W., Eds.; CRC Press/Balkema: London, UK, 2015; pp. 2685–2691.
18. Kumar, S.; Jakhar, M. Understanding user evaluation of Information Quality Dimensions in a digitized world. In Proceedings of the POMS Conference, Orlando, FL, USA, 6–9 May 2016.
19. Arazy, O.; Kopak, R.; Hadar, I. Heuristic Principles and Differential Judgments in the Assessment of Information Quality. *JAIS* **2017**, *18*, 403–432. [CrossRef]
20. Lee, Y.W.; Strong, D.M.; Kahn, B.K.; Wang, R.Y. AIMQ: A methodology for information quality assessment. *Inf. Manag.* **2002**, *40*, 133–146. [CrossRef]
21. Eppler, M.; Muenzenmayer, P. Measuring Information Quality in the Web Context: A Survey of State-of-the-Art Instruments and an Application Methodology. In Proceedings of the Seventh International Conference on Information Quality (ICIQ '02), Boston, MA, USA, 8–10 November 2002; pp. 187–196.
22. Greal, G. The Framework of Quality Measurement. *Management* **2015**, *10*, 177–191.
23. Todoran, I.; Lecornu, L.; Khencha, f A.; Le Caillec, J. Information quality evaluation in fusion systems. In Proceedings of the 16th International Conference on Information Fusion, Istanbul, Turkey, 9–12 July 2013; pp. 906–913.
24. Keeton, K.; Mehra, P.; Wilkes, J. Do you know your IQ? A research agenda for information quality in systems. *SIGMETRICS Perform. Eval. Rev.* **2009**, *37*, 26–31. [CrossRef]
25. Lesot, MJ.; Revault d'Allonnes, A. Information Quality and Uncertainty. In *Uncertainty Modeling. Studies in Computational Intelligence*; Kreinovich, V., Ed.; Springer International Publishing: Cham, Switzerland, 2017; Volume 683.
26. Stawowy, M.; Rosiński, A.; Paś, J.; Klimczak, T. Method of Estimating Uncertainty as a Way to Evaluate Continuity Quality of Power Supply in Hospital Devices. *Energies* **2021**, *14*, 486. [CrossRef]
27. Stawowy, M. *Method of Multilayer Modeling of Uncertainty in Estimating the Information Quality of ICT Systems in Transport*; Publishing House Warsaw University of Technology: Warsaw, Poland, 2019.
28. Becerra, M.A.; Tobón, C.; Castro-Ospina, A.E.; Peluffo-Ordóñez, D.H. Information Quality Assessment for Data Fusion Systems. *Data* **2021**, *6*, 60. [CrossRef]
29. Stawowy, M.; Rosiński, A.; Siergiejczyk, M.; Perlicki, K. Quality and Reliability-Exploitation Modeling of Power Supply Systems. *Energies* **2021**, *14*, 2727. [CrossRef]
30. Rychlicki, M.; Kasprzyk, Z.; Rosiński, A. Analysis of Accuracy and Reliability of Different Types of GPS. *Receiv. Sens.* **2020**, *20*, 6498. [CrossRef]
31. Klimczak, T.; Paś, J. *Basics of Exploitation of Fire Alarm Systems in Transport Facilities*; Military University of Technology: Warsaw, Poland, 2020.
32. Duer, S. Applications of an artificial intelligence for servicing of a technical object. *Neural Comput. Appl.* **2013**, *22*, 955–968. [CrossRef]
33. Bednarek, M.; Dąbrowski, T.; Olchowik, W. Selected practical aspects of communication diagnosis in the industrial network. *KONBiN* **2019**, *49*, 383–404. [CrossRef]
34. Szmel, D.; Zabłocki, W.; Ilczuk, P.; Kochan, A. Selected issues of risk assessment in relation to railway signalling systems. *WUT J. Transp. Eng.* **2019**, *127*, 81–91. [CrossRef]
35. Jacyna, M.; Żak, J.; Gołębiowski, P. The EMITRANSYS model and the possibilities of its application for the analysis of the development of sustainable transport systems. *Combust. Engines* **2019**, *179*, 243–248. [CrossRef]
36. Oleński, J. *Economics of Information. The Basics*; PWE: Warsaw, Poland, 2001.
37. Dempster, A.P. Upper and Lower Probabilities Inducted by a Multi-valued Mapping. *Ann. Math. Stat.* **1967**, *38*, 325–339. [CrossRef]
38. Shafer, G. *A Mathematical Theory of Evidence*; PUP: Princeton, NJ, USA, 1976.
39. Pieczynski, W. Unsupervised Dempster-Shafer Fusion of Dependent Sensors. In Proceedings of the 4th IEEE Southwest Symposium on Image Analysis and Interpretation, Austin, TX, USA, 2–4 April 2000; pp. 247–251.

Article

The Impact of Temperature of the Tripping Thresholds of Intrusion Detection System Detection Circuits

Jarosław Łukasiak [1], Adam Rosiński [2,*] and Michał Wiśnios [1]

[1]   Division of Electronic Systems Exploitations, Institute of Electronic Systems, Faculty of Electronics, Military University of Technology, 2 Gen. S. Kaliski St, 00-908 Warsaw, Poland; jaroslaw.lukasiak@wat.edu.pl (J.Ł.); michal.wisnios@wat.edu.pl (M.W.)
[2]   Division Telecommunications in Transport, Faculty of Transport, Warsaw University of Technology, 75 Koszykowa St, 00-662 Warsaw, Poland
*   Correspondence: adam.rosinski@pw.edu.pl

Abstract: This research paper discusses issues regarding the impact of temperature on the tripping thresholds of intrusion detection system detection circuits. The objective of conducted studies was the verification of a hypothesis assuming that the variability of an intrusion detection system's (considered as a whole) operating environment temperature can impact the electrical parameters of its detection circuits significantly enough so that it enables a change in the interpretation of the state observed within a given circuit fragment from the state of "no circuit violation" to "circuit violation". The research covered an intrusion detection system placed in a climatic chamber with adjusted temperature ($-25.1 \div +60.0$ [°C]). The analysis of the obtained results enabled determining the relationships that allow selecting detection circuit resistor values. It is important since it increases the safety level of protected facilities through proper resistor selection, thus, correct interpretation of a detection circuit state.

Keywords: intrusion detection system; detection circuit; temperature; tripping thresholds; climatic chamber; temperature characteristics; diagnosis reliability

Citation: Łukasiak, J.; Rosiński, A.; Wiśnios, M. The Impact of Temperature of the Tripping Thresholds of Intrusion Detection System Detection Circuits. *Energies* **2021**, *14*, 6851. https://doi.org/10.3390/en14206851

Academic Editor: Stanisław Duer

Received: 17 September 2021
Accepted: 16 October 2021
Published: 19 October 2021

Publisher's Note: MDPI stays neutral with regard to jurisdictional claims in published maps and institutional affiliations.

1. Introduction

The document National Critical Infrastructure Protection Program includes 11 systems as part of the critical infrastructure in the Republic of Poland. They are of crucial importance to the security of the state and its citizens. Their correct functioning ensures efficient operation of public administration authorities and the entrepreneurs. Critical infrastructure includes the following systems [1]:

- supply with power, power raw materials and fuels [2,3],
- communications,
- ICT networks,
- financial [4],
- food supply,
- water supply,
- health care,
- transport [5,6],
- emergency services,
- ensuring continuity of the public administration operation,
- production, deposition, storage and the use of chemicals and radioactive substances (including pipelines with hazardous substances).

One of the most important systems is transport. According to [1], it is the displacement of people, cargo (transport subject) in space, using appropriate means of transport. An efficient transport system is one of the pillars of a modern country [7]. Therefore, it is important to ensure security of objects (both stationary and mobile) participating in the transport system [8,9]. This is achieved by using, among others, electronic security systems [10,11].

Electronic security systems are composed of the following systems distinguished depending on the detected threats. These are:

- intrusion detection systems [12,13],
- access control systems [14,15],
- CCTV system [16],
- fire alarm systems [17–19].

The use of correctly designed alarm transmission systems [20–22] (including communication between vehicles and infrastructure [23]), which enables sending information from individual systems to an alarm receiving center [24] (including fire alarm system [25]) is also important.

The authors of this research paper discussed issues regarding the impact of temperature on the tripping thresholds of intrusion detection system detection circuits. This is an important issue in terms of protecting transport objects since they function under difficult environmental conditions [26–28], including the presence of high temperature variability [29]. Therefore, the issue of correctly defining, which state of the system can be deemed permissible or unacceptable from the point of view of security of an intrusion detection system (IDS) is important.

Wired alarm systems make the decision on recognizing an alarm based on analyzing a number of signals received via detection circuits from a wide range of sensors that can be used [30]. The most important examples of sensors include motion detectors (PIR-Passive Infra-Red and dual), magnetic, and others. The fact of them recognizing a factor classified as a threat (human presences, door and window opening, detection of gases harmful to human health and life) are signaled by a change of its electric parameters (most usually resistance), hence, the electric properties of the very detection circuit. The set of distinguishable threat states (e.g., alarm, sabotage, etc.) is defined by the number and manner of connection between EOL resistors and a given sensor. In consequence, the following configuration are distinguished: NC (Normally Closed), NO (Normally Open), EOL (End of Line), and 2EOL (alternatively DEOL—Double End of Line), which can also appear in NC and NO variants. From a practical perspective, alarm central units control the circuit state most usually through measuring the voltage drop across known resistances, which include EOL resistors, among others. A certain voltage range corresponds to each of them in a given configuration. The range can be translated to a proportional margin of permissible resistance values of the used EOL resistors.

The resistor, besides its natural feature—nominal resistance, is characterized by its tolerance coefficient, which results from the manufacturing spread. Their values are applied, in different form, virtually onto every currently manufactured resistor in the Through-Hole Technology (THT). Other important parameters, besides the aforementioned ones, include maximum power that can be generated within a characterized passive element, and also the often-forgotten temperature coefficient, which defines the nominal resistance power that can change per each temperature change of 1 [K].

The manufacturers of intrusion detection systems define their own preferred nominal values for EOL resistors. The most popular include: 1.1 [kΩ], 2.2 [kΩ], and 5.6 [kΩ]). It is not uncommon for a new IDS to be factory-fitted with a strip of several (usually approx. 15) aforementioned electronic elements in the layered technology. A photographic example of the said subassemblies is shown in Figure 1. Selected alarm systems permit the use of any EOL resistor values falling within a range specified by its manufacturer.

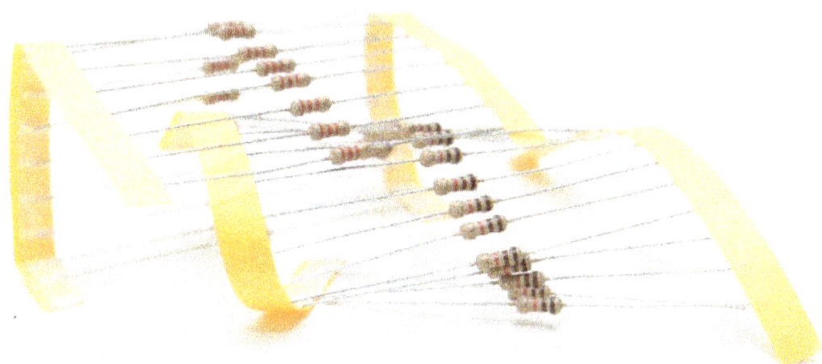

Figure 1. Typical layered EOL resistors fitted as equipment for intrusion detection systems (IDS).

2. Literature Review

The impact of temperature on the functioning of common-use electronic equipment digital systems was known to the authors of this article to have been previously discussed in publications. Such examples include [31], the authors of which discussed a temperature analysis involving the reliability of key electronic subassemblies on a PCB. This enabled optimizing the arrangement of electronic subassemblies based on temperature modelling using the Finite Element Method (FEM) using Ansys software. The developed method does not change the external cooling state, but leads to reduced maximum PCB temperature, thus improving reliability indicators.

A similar approach was presented in [32], with a proposed original temperature imaging platform dedicated to monitoring temperature distribution on the surfaces of PCBs in small electronic devices and systems. A thermal imaging system using the Arduino platform and an IR temperature sensor were used to this end. This makes the method inexpensive and accessible.

The authors of [33] also elaborated on the temperature distribution in electronic devices with natural air cooling. The conducted analyses enabled developing a mathematical model for mass and thermal characteristics, which contains equations defining the optimal number of printed-circuit boards, distances between the boards, and rail width. However, it does not take into account connections with peripherals.

An important issue when determining the impact of temperature on the functioning of a studied system is taking into account a relevant temperature sensor and test circuit selection. Such considerations are included in [34]. The authors described dynamic reliability tests involving the temperature properties of electronic subassemblies. The article focuses on suggesting a measuring system for dynamic high-temperature measurements of electronic systems. Particular attention was given to the issue of the temperature sensor and signal interference arising from the application of long measuring cables from the sensor to signal conditioning and processing devices. Temperature compensation was yet another aspect of the measurements. Most usually, a studied system or device, or its temperature to be more precise, is in reality higher than that of the measuring space, since the internal electronic device is also a heat source. This can lead to a measurement error. The authors of this article minimized such errors.

The issue associated with the impact of temperature on the functioning of electronic devices is very important in terms of the security of protected property and information. Discussions in this aspect were included in [35], which analyzed the possibility of a thermal attack of cryptographic devices and electronic modules. In order to protect logic circuits against functional errors due to operation in temperatures exceeding the operating range permitted by the manufacturer, the authors suggested a prototype of an active PCB tamper system with temperature monitoring. This is a solution beneficial to system arranged on a PCB, however; it does not protect detection circuits. Similar discussions were presented in the elaboration [36], with focus on PCBs themselves and the application of dedicated paths (so-called conductive mesh). The issues associated with a thermal attack on electronic devices of security systems are significant, which is why the authors of this article focused on determining the impact of temperature on the tripping thresholds of IDS detection circuits.

Another important problem in ensuring adequate reliability of electronic systems is the application of appropriate solder. Deliberations in this area are included in [37], the authors of which conducted low-temperature reliability tests of lead solder. Such solder is used in specialized electronic devices (industrial, military, medical, aeronautics), whereas other consumer electronics utilizes lead-free solder. The publication [38] also contains discussions on temperature fatigue testing of the elements installed in the Package-on-Package (PoP) technology. Reliability was determined through monitoring resistance for test PCBs placed in a climatic chamber.

The work [39] describes climatic resistance testing of heating layers integrated in PCBs during climatic tests. The study involved climatic tests of trial PCBs with integrated "underfloor heating" aimed at preventing condensation on soldered electronic systems under specific climatic conditions. The authors conducted experiments in a climatic chamber also in the case of determining the impact of temperature on the tripping thresholds of IDS detection circuits.

The issue of ensuring adequate reliability of electronic devices already at the engineering stage is approached in numerous studies. Ref. [40] presents a method for increasing reliability of electronic subassemblies through analyzing electric, temperature and mechanical load reserves at an early engineering stage of the electronic device, based on ASONIK software. It is important to take into account maximum permissible temperatures, and vibration and shock accelerations in electronic components.

The aspects in terms of the quality of information [41] transmitted from to the device from sensors are also crucial when analyzing the impact of temperature on the tripping thresholds of IDS detection circuits. Artificial neural networks are used in some scientific studies regarding reliability and operation [42,43]. The functioning of an intrusion detection system detection circuit is also impacted by vibrations [44–46] but they were not included in the tests covered by this article.

Despite so many works in the field of the impact of temperature on the functioning of digital systems, there are no deliberations directly addressing the impact of temperature on the tripping thresholds of IDS detection circuits. For this purpose, the authors of this article conducted tests in this regard.

Electronic security systems, due to their particular objectives they must fulfil and the entailing high responsibility, must be characterized not only by high reliability in terms of the hardware [47–49], but also understood as a decision on the security of a monitored area, created based on data collected previously from the aforementioned sensors. It should be noted that, in the context of the described studies, especially the second of the aforementioned aspect may be considered as a diagnostic and measurement issue. Monitoring the voltage drop across a specific reference resistance, which includes, among others, EOL (parametric) resistors determines a diagnosis in one of the following forms—violated detection circuit, non-violated detection circuits, and sabotaged detection circuit. For obvious reasons, the manufacturers of the analyzed group of electronic systems do not publish information on the reliability and efficiency of their solutions. On one hand, this is determined by the desire to protect their intrusion detection systems against the disclosure of their potential vulnerabilities and suggesting potential intruders to concentrate their efforts on other areas. Whereas from the perspective of people and property monitored by an IDS, it seems to be a non-transparent approach, which forces a consumer to entrust his/her property, and also health, in extreme cases, solely based on manufacturer's assurances regarding the effectiveness of the offered system, which may or may not be supported by arguments, verified and confirmed via experiments.

3. Materials and Methods

The baseline test configuration of a detection circuit within the conducted experiments was EOL, the structure of which is shown in Figure 2. This decision was determined by the fact that it is the simplest topology, which includes EOL resistors, resulting unfortunately in a lower number of distinguished detection circuit states, compared to the 2EOL variant [30].

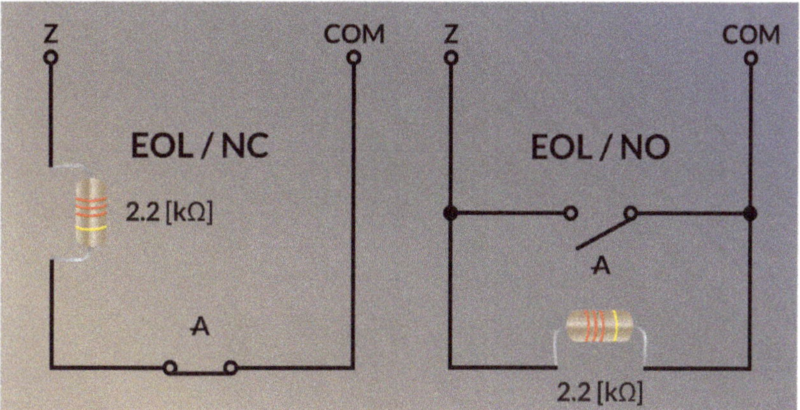

Figure 2. Structure of a parametric detection circuit of and EOL-type NC and NO IDS.

The experimental system used was an Integra 64 control panel, hardware version 1.4 B, by a Gdańsk-based manufacturer—Satel sp. z o.o. Terminal blocks (also known as KEFA connectors) located on the PCB (Printed Circuit Board) of the IDS were coupled with, via 2.6 m (8.53 [ft]) long category 5 Alcatel Data Cable UTP flex 4PR patchable 7x.07 network cable, an INT-KLCD-GR keypad, in accordance with the guidelines in the manufacturer's manual, taking into account the instruction for the clock signal and data line not to be routed via cables within the same twisted pair [30]. The keypad was fitted with a hardware programming interface marked USB-RS. According to the manufacturer's guidelines, the unused programmable HV outputs of the control panel were loaded with 2.2 [kΩ] resistors included in the set. The main supply path was fed by a TRZ 50/20 transformer by Pulsar sp. j. No additional power source in the form of a maintenance-free gel battery was used. Physical terminals of the first detection circuit, via a ca. 2.58 m (8.46 [ft]) long cat. 5 AT&T-I Systimax 1061c 4/24 cm cable, with a universal multi-contact PCB. In the course of the tests, it was coupled with a factory-ready EOL resistor with a resistance of 2.2 [kΩ] and a tolerance of 5% or a hard-wired, precise, 10-rotation WXD3-13-2W potentiometer by Chengdu Guosheng Technology Co., Ltd., with a nominal value of 4.7 [kΩ], 5% tolerance, and rated power of 2 [W]. The physical connectors of the second detection circuit were directly fitted with a factory-supplied layered EOL resistor with a value of 2.2 [kΩ] and a 5% tolerance.

In addition to the aforementioned elements, the authors prepared a 2.67 m (8.76 [ft]) section of a cat. 5 AT&T-I Systimax 1061c 4/24 cm network cable. The two subsequent pairs of which had soldered factory-supplied layered EOL resistors with a tolerance of 5%, and nominal values of 2.2 [kΩ] and 1.1 [kΩ], respectively.

The motherboard of the intrusion detection system subject to experiments, fastened together with a transformer to a housing, was placed in a Lab Event L C/100/70/10 climatic chamber by Weiss Technik GmbH. A system keyboard with a communicating programming interface, detection circuit coupled with a universal multi-contact board and a network cable with soldered test EOL resistors, as well as an IDS supply cable were routed out of the chamber via temperature-tight technical bushings. The programming interface was also coupled to a computer with installed GuardX software of the manufacturer, intended for alarm system management [30]. The view and a block diagram of the test bench was presented in Figures 3 and 4.

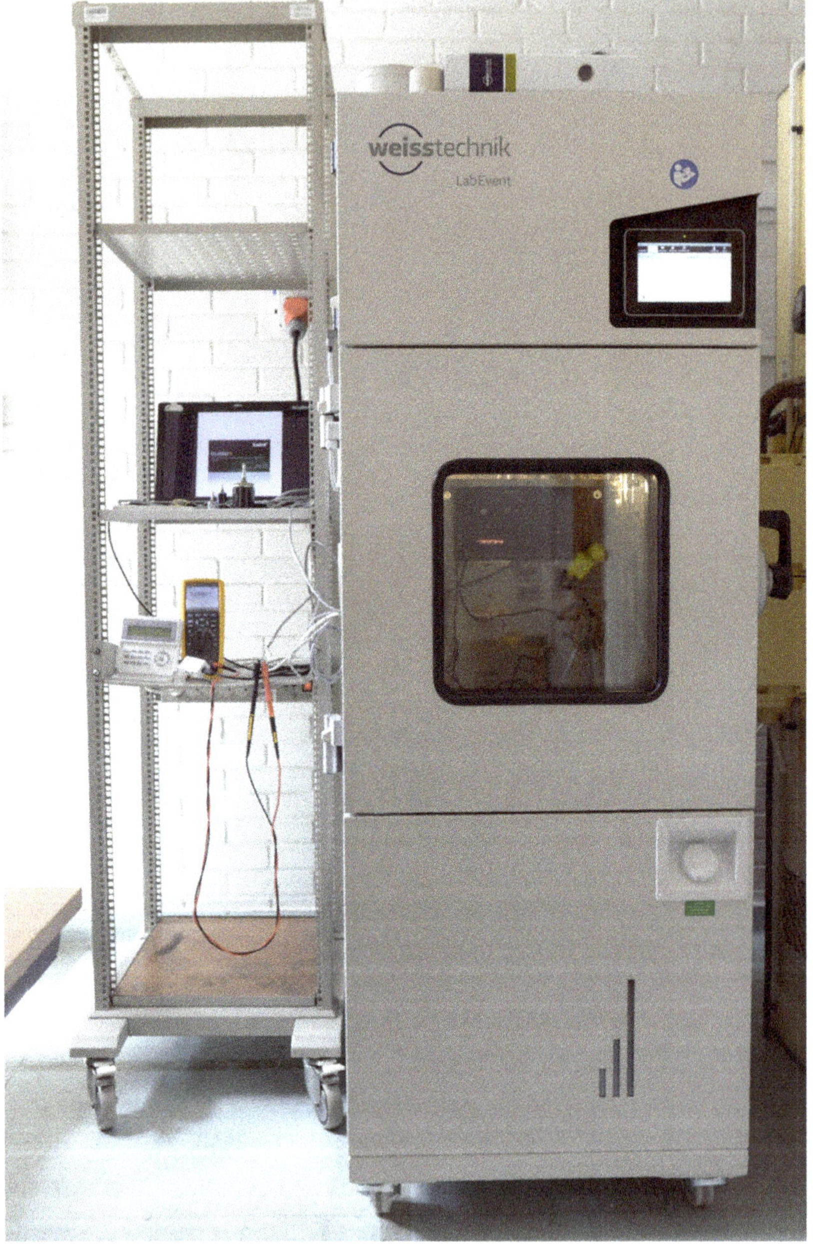

Figure 3. Test bench for studying the impact of temperature on the tripping thresholds of intrusion detection system detection circuits.

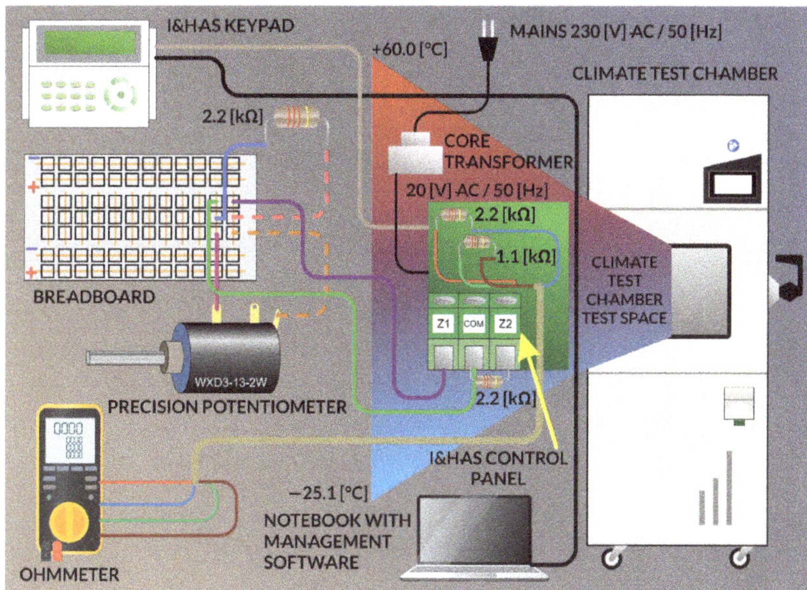

Figure 4. Block diagram of a test bench for studying temperature characteristics of layered EOL resistors and the impact of temperature on the resistance thresholds that determine the security states of IDS detection circuits.

The measurements were taken as per the following methodology [50–55]. A desired temperature in the non-stop operating mode was entered on the control panel of the climatic chamber. The aforementioned function results in the device prioritizing reaching and stabilizing the preset temperature in the measuring space as fast as possible, while omitting the humidity parameter and its potential variability [50]. Next, the researchers waited until the temperature inside the chamber stabilized, with the least possible deviation. It should be noted that the available equipment, besides striving to achieve declared temperature, also ensures compensation of the heat generated by the DUT (Device Under Test) located within the measuring space. For this reason, the authors discarded connecting a backup power source in order not to unnecessarily burden the IDS control panel with the need to load a battery, which would lead to intensified heat dissipation, resulting in the chamber having to taken on its additional compensation. Summing up, the tested intrusion detection system was to operate based on the simplest possible configuration, so that the emitted heat was the lowest, which should shorten the time of reaching and stabilizing the preset temperature by the climatic chamber.

The next step involved measuring the resistance of two layered EOL resistors located in the temperature-stabilized measuring space using a Fluke 289 digital multimeter.

Next, the universal contact board was coupled with a detection circuit and a layered EOL resistor with a nominal value of 2.2 [kΩ] and a 5% tolerance. The computer run GuardX software that can be used to monitor and visualize security states of individual circuits within a tested IDS in real time. No violation of the aforementioned system fragment is depicted by a square filled with grey color. The violation (detection circuit resistance above or below the threshold value) is announced by a change in the box color to green. When the violation time exceeded a set limit, the application signaled the described event with information on failure due to prolonged violation and through alternatively changing the color of the square from grey to orange. Characterized situations are shown in Figure 5. The presence of an EOL resistor on the contact board is equivalent to lack of detection zone violation. Removing this element led to the intrusion detection system interpreting this phenomenon as a circuit violation. Remounting the subassembly into the universal contact board resulting in changing the detection zone status from "no violation" both after violation and when the resistor was installed after a prolonged period of time that was announced as a failure situation.

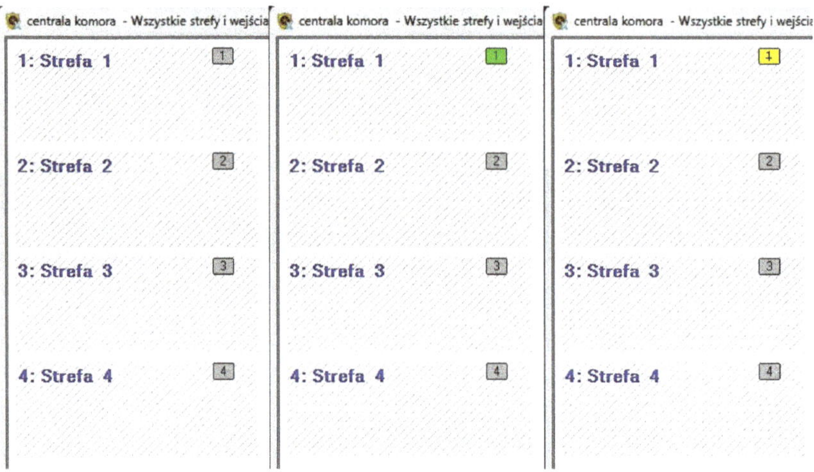

Figure 5. Visualization of the states—no violation, violation, long violation visualized via GuardX software for managing Satel alarm control panels.

The course of the experiment involved repeated use of a Fluke 289 digital multimeter to determine and then record a specific resistance of the multi-turn potentiometer uncoupled from the circuit. Next, the EOL resistor was removed from the multi-contact board, which was immediately communicated as detection circuit violation. A potentiometer with a preset resistance value was used to replace the resistor, remembering about preceding this operation with uncoupling the ohmmeter. The response of the system was then observed and the conclusions recorded. The verification of IDS decision on the state of a detection circuit for each of the potentiometer setting at a preset operating temperature point was checked four times. The entire described sequence of operations was repeated after changing the operating temperature point of the climatic chamber measuring space.

In the event of declaring a detection circuit as EOL, the programming app for Satel systems (DloadX) does not distinguish between the NC and NO variants [30]. This is due to the use of upper and lower resistance thresholds, between which there is a manufacturer-determined parametric value (2.2 [kΩ]). In such a situation, exceeding the limit value determined by the upper threshold, is classified from the perspective of the control panel as infinite resistance, which corresponds to opening of the relay due to detecting a phenomenon constituting a threat in the NC variant. Whereas exceeding the lower parametric resistance threshold is diagnosed by the IDS as a short-circuit, which corresponds to a violation of the EOL NO detection circuit. An intermediate objective of the aforementioned measurements is determining the lower and upper resistance thresholds activating the parametric detection circuits within the EOL configuration.

Studying the temperature characteristics of layered EOL resistors included by the manufacturer will enable estimating the values of their temperature coefficients.

4. Results

The primary objective of the conducted experiments was to simulate extremely adverse atmospheric operating conditions for an electronic security system represented by an intrusion detection system, studying their impact on EOL resistors and the potential shift of the upper and lower parametric resistance thresholds, which when exceeded would trip an IDS alarm. These test results enabled unequivocal determination whether, assuming the overlapping of extremely adverse weather conditions, the event of a false detection circuit violation will be possible in the case of a system functioning at a minimum permissible operating temperature, with installed EOL resistor exhibiting a value at the border of the manufacturing spread, and taking into account potential displacement of parametric system trip thresholds. Similar deliberations should be related to the second boundary condition, hence, the maximum permissible operating temperature for the studied IDS. The aforementioned range for Integra control panels was specified at $-10 \div +55$ [°C]. Due to the specific significance of electronic security systems, it was decided to expand the range of test temperature in the climatic chamber to $-25.1 \div +60.0$ [°C].

The results of measuring the characteristics of layered EOL resistors are listed in Table 1. They are graphically presented in Figure 6. The result analysis clearly indicates a linear dependence of the change in the resistance of the aforementioned passive elements, as a function of temperature. This can be used as a basis to estimate the temperature coefficient for the resistors subject to testing. In the case of the resistor in question, with the nominal value of 1.1 [kΩ], the aforementioned parameter ranges from approx. 249 to 268 [ppm/K], whereas for a resistor with the nominal value of 2.2 [kΩ], this parameter falls in the range of approx. $482 \div 529$ [ppm/K]. This enables a conclusion that in the case of both analyzed EOL resistors, their temperature coefficient value intervals fall within the range of values typical for layered resistors.

Table 1. Results for the measurement of temperature characteristics of EOL resistors with the nominal values of 1.1 [kΩ] and 2.2 [kΩ].

Change in the resistance of a resistor with a nominal value of 1.1 [kΩ] and a 5% tolerance, as a function of temperature					
Resistance [kΩ]	1.0985	1.1034	1.1075	1.1168	1.1207
Temperature [°C]	60.0	40.3	25.0	−10.1	−25.1
Change of resistance from $T_{min.}$ to $T_{max.}$ [kΩ]			0.0222		
Change of resistance from 25.1 [°C] to 60.0 [°C]			0.0090		
Change of resistance from 25.0 [°C] to −25.1 [°C]			0.0132		
Change in the resistance of a resistor with a nominal value of 2.2 [kΩ] and a 5% tolerance, as a function of temperature					
Resistance [kΩ]	2.1730	2.1825	2.1906	2.2089	2.2167
Temperature [°C]	60.0	40.3	25.0	−10.1	−25.1
Change of resistance from $T_{min.}$ to $T_{max.}$ [kΩ]			0.0437		
Change of resistance from 25.1 [°C] to 60.0 [°C]			0.0176		
Change of resistance from 25.0 [°C] to −25.1 [°C]			0.0261		

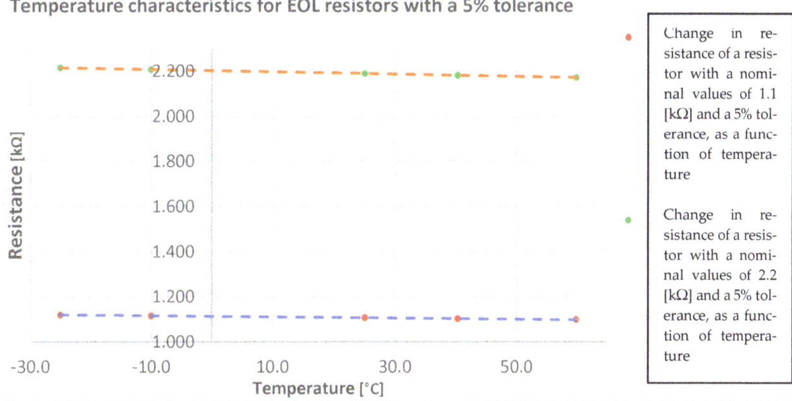

Figure 6. Temperature characteristics with EOL resistors with a tolerance of 5%.

Continuing the discussion, it might be worthwhile to determine the change in the resistance of individual EOL resistors upon significant temperature deviations, relative to room temperature (25.0 [°C]), adopted as the reference point. Further analysis of the conducted experiment results indicates that a resistor with the nominal value of 1.1 [kΩ], upon a temperature change in the range of +25.0 ÷ +60.0 [°C], changes the value of its primary parameter by 0.0090 [Ω] and by 0.0132 [Ω], for a temperature range of +25.0 ÷ −25.1 [°C]. In the case of an EOL resistor with the nominal value of 2.2 [kΩ], and at similar temperature ranges, resistance changes equal to 0.0176 [Ω] and 0.0261 [Ω], respectively. Table 1 lists the results of the analysis above.

Given the 5% manufacturing spread in the studied resistors, the range of statistically achievable real resistances of the resistors with the nominal values of 1.1 [kΩ] and 2.2 [kΩ] can be estimated. The aforementioned ranges are summarized in Table 2.

Table 2. Permissible ranges of the resistance reached by real resistors with the nominal values of 1.1 [kΩ] and 2.2 [kΩ], at a manufacturing spread of 5%.

	$R_{min.}$ [Ω]	$R_{max.}$ [Ω]
Change in the resistance of a resistor with a nominal value of 1.1 [kΩ] resulting from manufacturing spread	1045	1155
	$R_{min.}$ [Ω]	$R_{max.}$ [Ω]
Change in the resistance of a resistor with a nominal value of 2.2 [kΩ] resulting from manufacturing spread	2090	2310

Assuming the superpositions of extremely adverse circumstances related to boundary cases for all previous deliberations, it is possible to determine the ranges for the variability of nominal resistance in the tested EOL resistors. The aforementioned data is shown in Table 3.

Table 3. Permissible ranges of the resistance achieved by real EOL resistors, taking into account the superposition of most unfavorable circumstances.

Resistor with a nominal value of 1.1 [kΩ]		
Minimum and maximum resistance resulting from the minimum and maximum superpositions of manufacturing spread resistance and resistance changes at a minimum and maximum temperature	1.0318	1.1640
Resistor with a nominal value of 2.2 [kΩ]		
Minimum and maximum resistance resulting from the minimum and maximum superpositions of manufacturing spread resistance and resistance changes at a minimum and maximum temperature	2.0639	2.3276

In order to be able to clearly state whether the aforementioned extreme cases can have a significant impact on the ultimate decision of the intrusion detection system regarding the diagnosis on the state of the detection circuit, the results from Table 3 should be compared with the results of measurements aimed at determining the threshold values for parametric resistances, which determine the change of the detection circuit state from "no violation" to "violated". Characterized test outcomes are listed in Table 4, while for the sake of result presentation clarity, the "detection circuit violated" state has been assigned the red color. The aforementioned designation was assigned to results that were signaled by GuardX software as a too long violation for all measurements at a given operating temperature points. Green color was allocated to the results that lead to a constant message on the lack of detection circuit violation for all four test trials. Orange was assigned to ambiguous results, i.e., those which were composed of any combination of the aforementioned states in four measurements at a given operating temperature point, with a specified setting of the multi-turn precise potentiometer. It should be noted that despite the availability of a very advanced climatic chamber, the authors experienced temperature deviations relative to the values preset on the control panel, the range of which is included in Table 4. The system in question, just like every other electronic device, transforms some of the electric power consumed for operating purposes it was designed to and to process data, into heat as the loss power. In such a case, the task of the climatic chamber was not only to maintain a constant temperature within its measuring space, but also to provide follow-up compensation of the thermal energy generated by the system under test.

Table 4. Resistance measurement results for intrusion detection system threshold tripping limits, as a function of temperature.

Temperature: 60.0 [°C]		Temperature: 40.2 ± 0.1 [°C]		Temperature: 25.0 ± 0.2 [°C]		Temperature: −25.1 ± 0.1 [°C]	
Resistance [kΩ]	Parametric Line State	Resistance [kΩ]	Parametric Line State	Resistance [kΩ]	Parametric Line State	Resistance [kΩ]	Parametric Line State
2.8547	A, A, A, A	2.8549	A, A, A, A	2.8549	A, A, A, A	2.8554	A, A, A, A
2.8535	A, A, A, A	2.8534	A, A, A, A	2.8533	A, A, A, A	2.8533	A, A, A, A
2.8524	A, A, A, A	2.8525	A, A, A, A	2.8528	A, A, A, A	2.8528	A, A, A, A
2.8512	A, A, A, NA	2.8515	A, A, A, A	2.8516	A, A, A, A	2.8514	A, A, A, A
2.8496	A, A, NA, A	2.8498	A, A, A, A	2.8503	A, A, A, A	2.8501	A, A, A, A
2.8485	A, A, A, A	2.8484	A, A, A, A	2.8484	A, A, A, A	2.8486	A, A, A, A
2.8471	A, NA, NA, NA	2.8472	A, A, A, A	2.8476	A, A, A, A	2.8476	A, A, A, A
2.8457	A, NA, NA, A	2.8459	A, A, A, A	2.8463	A, A, A, A	2.8462	A, A, A, A
2.8444	NA, NA, NA, NA	2.8446	A, A, A, A	2.8449	A, A, A, A	2.8449	A, A, A, A
2.8432	NA, NA, NA, NA	2.8435	A, A, A, A	2.8436	A, NA, A, A	2.8436	A, A, A, A
2.8418	NA, NA, NA, NA	2.8421	NA, A, A, NA	2.8423	A, A, A, A	2.8422	A, A, A, A
2.8406	NA, NA, NA, NA	2.8408	A, NA, NA, NA	2.8410	A, A, A, A	2.8409	A, A, A, A
2.8391	NA, NA, NA, NA	2.8394	NA, NA, NA, NA	2.8391	A, A, NA, A	2.8396	A, A, A, A
2.8379	NA, NA, NA, NA	2.8379	NA, NA, NA, NA	2.8383	NA, NA, NA, NA	2.8383	A, A, A, A
2.8369	NA, NA, NA, NA	2.8368	NA, NA, NA, NA	2.8369	NA, A, NA, NA	2.8369	A, A, A, A
2.8352	NA, NA, NA, NA	2.8355	NA, NA, NA, NA	2.8357	NA, NA, NA, NA	2.8356	A, A, A, A
2.8339	NA, NA, NA, NA	2.8341	NA, NA, NA, NA	2.8343	NA, NA, NA, NA	2.8342	NA, A, A, A
2.8326	NA, NA, NA, NA	2.8329	NA, NA, NA, NA	2.8330	NA, NA, NA, NA	2.8330	A, NA, A, NA
2.8313	NA, NA, NA, NA	2.8316	NA, NA, NA, NA	2.8317	NA, NA, NA, NA	2.8317	A, A, NA, NA
2.8301	NA, NA, NA, NA	2.8303	NA, NA, NA, NA	2.8304	NA, NA, NA, NA	2.8305	NA, NA, NA, NA
2.8286	NA, NA, NA, NA	2.8282	NA, NA, NA, NA	2.8285	NA, NA, NA, NA	2.8290	NA, NA, NA, NA
2.8274	NA, NA, NA, NA	2.8277	NA, NA, NA, NA	2.8279	NA, NA, NA, NA	2.8278	NA, NA, NA, NA
2.8261	NA, NA, NA, NA	2.8263	NA, NA, NA, NA	2.8264	NA, NA, NA, NA	2.8264	NA, NA, NA, NA
2.8248	NA, NA, NA, NA	2.8246	NA, NA, NA, NA	2.8245	NA, NA, NA, NA	2.8237	NA, NA, NA, NA
⋮	⋮	⋮	⋮	⋮	⋮	⋮	⋮
2.2	NA, NA, NA, NA	2.2	NA, NA, NA, NA	2.2	NA, NA, NA, NA	2.2	NA, NA, NA, NA
⋮	⋮	⋮	⋮	⋮	⋮	⋮	⋮
1.5395	NA, NA, NA, NA	1.5397	NA, NA, NA, NA	1.5398	NA, NA, NA, NA	1.5397	NA, NA, NA, NA
1.5383	NA, NA, NA, NA	1.5384	NA, NA, NA, NA	1.5386	NA, NA, NA, NA	1.5385	NA, NA, NA, NA
1.5370	NA, NA, NA, NA	1.5371	NA, NA, NA, NA	1.5372	NA, NA, NA, NA	1.5372	NA, NA, NA, NA
1.5358	NA, NA, NA, NA	1.5359	NA, NA, NA, NA	1.5360	NA, NA, NA, NA	1.5359	NA, NA, NA, NA
1.5343	NA, NA, NA, NA	1.5345	NA, NA, NA, NA	1.5345	NA, NA, NA, NA	1.5345	NA, NA, NA, NA
1.5330	NA, NA, NA, NA	1.5332	NA, NA, NA, NA	1.5333	NA, NA, NA, NA	1.5333	NA, NA, NA, NA
1.5317	NA, NA, NA, NA	1.5319	NA, NA, NA, NA	1.5320	NA, NA, NA, NA	1.5320	NA, NA, NA, NA
1.5303	NA, NA, NA, NA	1.5306	NA, NA, NA, NA	1.5307	NA, NA, NA, NA	1.5307	NA, NA, NA, NA
1.5292	NA, A, NA, A	1.5293	NA, NA, NA, NA	1.5294	NA, NA, NA, NA	1.5294	NA, NA, NA, NA
1.5280	A, A, A, A	1.5280	NA, NA, NA, NA	1.5281	NA, NA, NA, NA	1.5281	NA, NA, NA, NA
1.5278	NA, A, A, A	1.5279	NA, NA, NA, NA	1.5280	NA, NA, NA, NA	1.5280	NA, NA, NA, NA
1.5266	A, A, A, A	1.5267	A, A, A, NA	1.5267	NA, NA, NA, NA	1.5267	NA, NA, NA, NA
1.5251	A, A, A, A	1.5255	A, A, A, A	1.5255	NA, NA, NA, A	1.5255	NA, NA, NA, NA
1.5239	A, A, A, A	1.5241	A, A, A, A	1.5241	A, A, A, A	1.5241	A, A, A, A
1.5226	A, A, A, A	1.5229	A, A, A, A	1.5229	A, A, A, NA	1.5228	A, A, A, A
1.5214	A, A, A, A	1.5215	A, A, A, A	1.5216	A, A, A, A	1.5214	A, A, A, A
1.5202	A, A, A, A	1.5203	A, A, A, A	1.5203	A, A, A, A	1.5202	A, A, A, A
1.5188	A, A, A, A	1.5189	A, A, A, A	1.5189	A, A, A, A	1.5189	A, A, A, A
				1.5177	A, A, A, A	1.5177	A, A, A, A

Marking in the Table 4: A-Alarm, NA-No Alarm.

When analyzing raw measurement data from Table 4 and dividing the sets of obtained detection circuit state by color into no violation (green), uncertain state (orange), and certain alarm state (red), one can almost immediately observe that the resistance determining the upper and lower detection circuit tripping thresholds grow as a temperature function. The last resistances corresponding to the non-violation state of a detection circuit, preceded solely by measurement points representing a state of certain non-violation were adopted as the limit values. For this reason, the result of 2.8383 [kΩ] for a temperature of 25.0 ± 0.2 [°C] could not be considered as the detection circuit upper tripping threshold.

A graphical interpretation of the characterized data is shown in Figures 7 and 8. Only then, one can notice a non-linear relationship between the values in question. The accuracy of conducted measurement can be proved by an almost unchanged bandwidth of the resistance values corresponding to the state of absolute detection circuit non-violation (difference between the obtained upper and lower thresholds for each test temperatures), which is approx. 1.31 [kΩ] for the range of +60.0 $\div$ −25.00 [°C], and decreases only by 9.1 [Ω]. Assuming the resistance of 2.2 [kΩ] as a point that should mark the middle of the band of values corresponding to the absolute detection circuit non-violation state, and based on the obtained results, it should be concluded that the aforementioned ranges are not symmetrical, starting already at room temperature. In the case of the measurement series in question, the 2.2 [kΩ] resistance and the upper threshold are separated by approx. 636 [Ω]. In the case of the lower threshold and the reference value, this range is equal to 673 [Ω]. Moreover, the observed disproportion worsens along with dropping temperature.

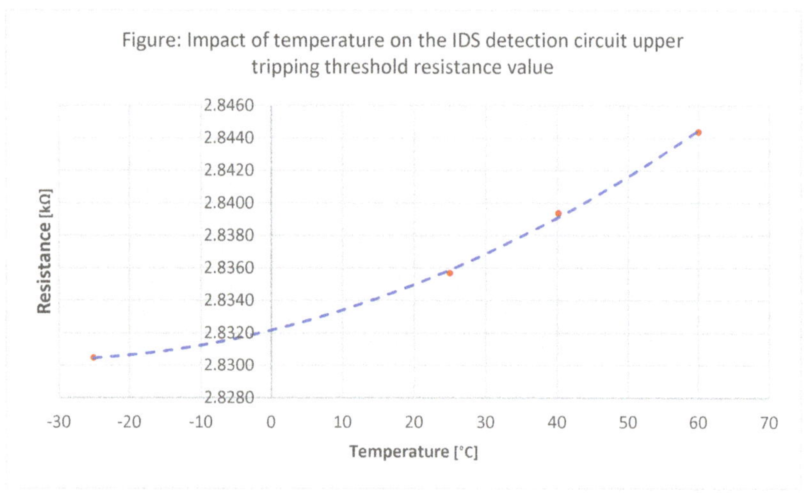

Figure 7. Impact of temperature of a parametric IDS circuit upper tripping threshold resistance value.

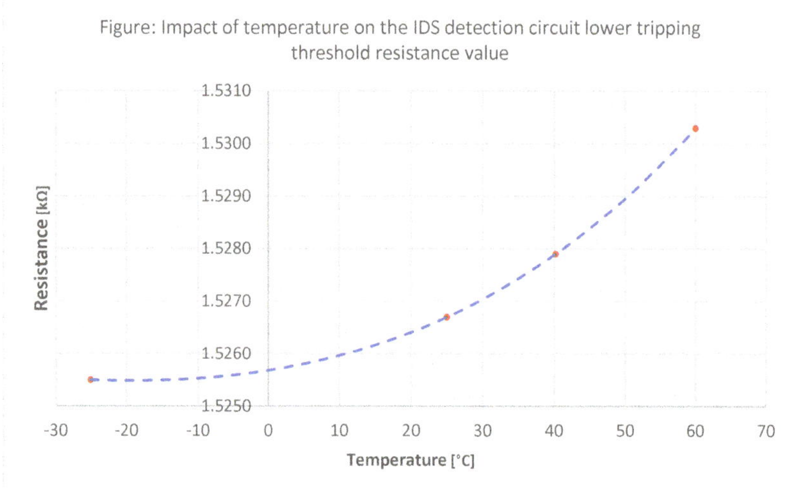

Figure 8. Impact of temperature of a parametric IDS circuit lower tripping threshold resistance value.

It should be noted that prior to conducting the primary tests, the authors also performed preliminary testing based on the 8143R10KL.25 precise multi-turn potentiometer by TT Electronics. The obtained results were similar with all of the presented above. The universal multi-contact board and elements used within the experiments are shown in Figure 9. In order to achieve better accuracy, the used adjusting element was changed to one ensuring better measurement resolution (instead of 10 [kΩ] per ten rotations to 4.7 [kΩ] per ten rotations).

Figure 9. The subassemblies used in the experiment as EOL resistors of an EOL detection circuit in an intrusion detection system.

The obtained upper and lower limit resistance thresholds allow for a conclusion that maintaining an adequate, safe resistance margin enables installers of the tested IDS to use EOL resistors with a value different than the one declared by the manufacturer. The author of this study, due to the range of values corresponding to the absolute detection circuit state of non-violation, compared to the factory-set value of 2.2 [kΩ] suggested two relationships, which enable determining the maximum and minimum value of a resistor, installing within a circuit will allow an intrusion detection system to function without impacting the diagnostic reliability in terms of the state within a specific, real detection circuit. In the case of the maximum resistors value, it was assumed that the installed passive element would be characterized by a 5 [%] tolerance. The proposed formula also takes into account the resistivity of a cable in a detection circuit, the total value of which has to be determined by the installed alone, and has to be used based on a catalogue card provided with the cabling, combined with the knowledge on the length of the created detection circuit. It was also assumed that the maximum value of the resistor used within a given tolerance cannot exceed the value separated from the lowest of them, and obtained via measurements of the upper threshold value (i.e., 2.8305 [kΩ] at −25.1 [°C]) with a 25% safety margin. The aforementioned consideration led to a relationship (1)

$$1.05 * x + y = 0.75 * 2.8305, \tag{1}$$

after transformations we get a Formula (2) for the maximum value of resistance suitable for practical application within a specific detection circuit of an intrusion detection system, expressed in [kΩ],

$$x = \frac{2.122875 - y}{1.05} \; [k\Omega], \tag{2}$$

where:

y-resistivity of the cable making up the detection circuit (please remember to take into account both the section from the IDS control panel to the resistor in the sensor and about the same-length cable coupled in the opposite direction).

Similar considerations apply to the minimum resistor value suitable for application as an EOL resistors, with the difference in that the use of the aforementioned passive element with a value lower than critical cannot exceed 25% of the security margin, relative to the highest lower threshold value (resulting in the narrowest range of resistance values, relative to the one preset by the manufacturer—2.2 [kΩ]). In this case, this was a value of 1.5303 [kΩ] obtained at a temperature of +60.0 [$^\circ$C]. Given the aforementioned assumptions, the following formula was obtained

$$0.95 * x + y = 0.75 * 1.5303, \tag{3}$$

which after transformation provides an equation,

$$x = \frac{1.147725 - y}{0.95} \ [k\Omega], \tag{4}$$

where:

y-resistivity of the cable making up the detection circuit (please remember to take into account both the section from the IDS control panel to the resistor in the sensor and about the same-length cable coupled in the opposite direction).

It should be stressed that the adopted 25% safety margin is aimed at including the temperature coefficient of the EOL resistor and a change in the resistivity of cables comprising a detection circuit due to ambient temperature changes.

5. Discussion and Conclusions

Given the obtained results of the experiment involving the impact of IDS operating environment temperature on tripping thresholds of individual detection circuit states and taking into account the measured temperature characteristics of real EOL resistors, it should be clearly concluded that there is no risk of a false alarm within the analyzed solution, caused by the aforementioned factor. The resistance variability in the case of considered passive elements, also given the range of temperature significantly exceeding typical environmental conditions specified by the manufacturer of the tested solution, is so small that it is impossible to approach typical thresholds changing individual detection circuit states (adopted as limit resistance values at room temperature). It should be stated that taking into account the observed minor deviations of limit resistances as a function of temperature will also not influence the reliability of distinguishing between the states of an IDS detection circuit. The deliberations prove that, assuming the case with the superposition of the most adverse conditions (i.e., overlapping of limit resistance threshold shift at an extreme temperature, causing the highest resistance changes, possible highest change in the resistance of the attached EOL resistor induced by the aforementioned extreme temperature, and the installation of a studied passive element, the value of which is a limit case of resistance falling within the 5% manufacturing spread determined by the manufacturer) will also not cause a misinterpretation of the current state of the detection circuit state.

Naturally, it should be noted that although the obtained results clearly indicate the correctness of the studied IDS design and the lack of a need to repeat the tests for other detection circuit configurations (e.g., 2EOL), it does not exclude the justification of repeating similar experiments not only for alarm control panel families of the same manufacturer (to a lesser extent), but above all, for solutions of their competitors. In this case aimed at detecting potential design errors, which involve the failure to predict or assuming too narrow resistance ranges, corresponding to individual states of detection circuits, which will naturally translate to very narrow voltage drop ranges.

Author Contributions: Conceptualization, J.Ł., A.R. and M.W.; Methodology, J.Ł. and M.W.; Validation, J.Ł. and A.R.; Formal analysis, J.Ł. and M.W.; Investigation, J.Ł. and A.R.; Resources, J.Ł. and M.W.; Data curation, J.Ł.; Writing—original draft preparation, J.Ł., A.R. and M.W.; Writing—review and editing, J.Ł., A.R. and M.W.; Visualization, J.Ł.; Supervision, A.R.; Project administration, A.R.; Funding acquisition, A.R. All authors have read and agreed to the published version of the manuscript.

Funding: This research received no external funding.

Institutional Review Board Statement: Not applicable.

Informed Consent Statement: Not applicable.

Data Availability Statement: The data presented in this study are available on request from the corresponding author.

Conflicts of Interest: The authors declare no conflict of interest.

References

1. Government Security Center. National Critical Infrastructure Protection Programme in Poland. August 2020. Available online: https://www.gov.pl/attachment/ee334990-ec9c-42ab-ae12-477608d94eb1 (accessed on 18 August 2021).
2. An, J.; Mikhaylov, A.; Richter, U.H. Trade War Effects: Evidence from Sectors of Energy and Resources in Africa. *Heliyon* **2020**, *6*, e05693. [CrossRef]
3. An, J.; Mikhaylov, A. Russian energy projects in South Africa. *J. Energy South. Afr.* **2020**, *31*, 58–64. [CrossRef]
4. Mishina, V.Y.; Khomyakova, I.I. Dedollarization and settlements in national currencies: Eurasian and Latin American experience. *Vopr. Ekon.* **2020**, *9*, 61–79. [CrossRef]
5. Gołębiowski, P.; Jacyna, M.; Stańczak, A. The Assessment of Energy Efficiency versus Planning of Rail Freight Traffic: A Case Study on the Example of Poland. *Energies* **2021**, *14*, 5629. [CrossRef]
6. Jacyna, M.; Żochowska, R.; Sobota, A.; Wasiak, M. Scenario Analyses of Exhaust Emissions Reduction through the Introduction of Electric Vehicles into the City. *Energies* **2021**, *14*, 2030. [CrossRef]
7. Losurdo, F.; Dileo, I.; Siergiejczyk, M.; Krzykowska, K.; Krzykowski, M. Innovation in the ICT Infrastructure as a Key Factor in Enhancing Road Safety: A Multi-Sectoral Approach. In Proceedings of the 2017 25th International Conference on Systems Engineering (ICSEng), Las Vegas, NV, USA, 22–24 August 2017; Selvaray, H., Chmaj, G., Zydek, D., Eds.; The Institute of Electrical and Electronics Engineers, Inc.: Danvers, MA, USA, 2017; pp. 157–162. [CrossRef]
8. Kierzkowski, A.; Kisiel, T. Simulation model of security control system functioning: A case study of the Wroclaw Airport terminal. *J. Air Transport. Manag.* **2017**, *64*, 173–185. [CrossRef]
9. Siergiejczyk, M.; Paś, J.; Dudek, E. Reliability analysis of aerodrome's electronic security systems taking into account electromagnetic interferences. In Proceedings of the Safety and Reliability—Theory and Applications, Proceedings of the 27th European Safety and Reliability Conference (Esrel 2017), Portorož, Slovenia, 18–22 June 2017; Čepin, M., Briš, R., Eds.; CRC Press/Balkema: Schipholewh, The Netherlands, 2017; pp. 2285–2292. [CrossRef]
10. Fischer, R.J.; Halibozek, E.P.; Walters, D.C. *Introduction to Security*, 10th ed.; Butterworth-Heinemann: Oxford, UK, 2019. [CrossRef]
11. Purpura, P.P. *Security and Loss Prevention: An Introduction*; Butterworth-Heinemann: Oxford, UK, 2019. [CrossRef]
12. Valouch, J. Technical requirements for Electromagnetic Compatibility of Alarm Systems. *Int. J. Circuits Syst. Signal Process.* **2015**, *9*, 186–191. Available online: https://www.naun.org/main/NAUN/circuitssystemssignal/2015/a522005-196.pdf (accessed on 10 July 2021).
13. Urbancokova, H.; Valouch, J.; Adamek, M. Testing of an intrusion and hold-up systems for electromagnetic susceptibility—EFT/B. *Int. J. Circuits Syst. Signal Process.* **2015**, *9*, 40–46. Available online: https://www.naun.org/main/NAUN/circuitssystemssignal/2015/a122005-024.pdf (accessed on 10 July 2021).
14. Wiśnios, M.; Paś, J. The assessment of exploitation process of power for access control system. *E3S Web Conf.* **2017**, *19*, 01034. [CrossRef]
15. Wiśnios, M.; Dąbrowski, T.; Bednarek, M. The security increasing level method provided by biometric access control system. *Przegląd Elektrotechniczny* **2015**, *91*, 229–232. [CrossRef]
16. Paś, J.; Rosiński, A.; Białek, K. A reliability-operational analysis of a track-side CCTV cabinet taking into account interference. *Bull. Pol. Acad. Sci. Tech. Sci.* **2021**, *69*, e136747. [CrossRef]
17. Klimczak, T.; Paś, J. *Basics of Exploitation of Fire Alarm Systems in Transport Facilities*; Military University of Technology: Warsaw, Poland, 2020.
18. Jakubowski, K.; Paś, J. Determination of the performance parameters of selected electronic safety systems based on the process of their use in critical infrastructure facilities. *Przegląd Elektrotechniczny* **2021**, *97*. [CrossRef]
19. Paś, J.; Klimczak, T. Selected issues of the reliability and operational assessment of a fire alarm system. *Eksploat. I Niezawodn. Maint. Reliab.* **2019**, *21*, 553–561. [CrossRef]

20. Krzykowska-Piotrowska, K.; Dudek, E.; Siergiejczyk, M.; Rosiński, A.; Wawrzyński, W. Is Secure Communication in the R2I (Robot-to-Infrastructure) Model Possible? Identification of Threats. *Energies* **2021**, *14*, 4702. [CrossRef]
21. Jacyna, M.; Szczepański, E.; Izdebski, M.; Jasiński, S.; Maciejewski, M. Characteristics of event recorders in Automatic Train Control systems. *Arch. Transport* **2018**, *46*, 61–70. [CrossRef]
22. Polak, R.; Laskowski, D.; Matyszkiel, R.; Łubkowski, P.; Konieczny, Ł.; Burdzik, R. Optimizing the Data Flow in a Network Communication between Railway Nodes. In *Research Methods and Solutions to Current Transport Problems, Proceedings of the International Scientific Conference Transport of the 21st Century, Advances in Intelligent Systems and Computing, Ryn, Poland, 9–12 June 2019*; Siergiejczyk, M., Krzykowska, K., Eds.; Springer: Cham, Switzerland, 2020; Volume 1032, pp. 351–362. [CrossRef]
23. Kossakowski, D.; Krzykowska, K. Application of V2X Technology in Communication between Vehicles and Infrastructure in Chosen Area. In *Research Methods and Solutions to Current Transport Problems, Proceedings of the International Scientific Conference Transport of the 21st Century, Advances in Intelligent Systems and Computing, Ryn, Poland, 9–12 June 2019*; Siergiejczyk, M., Krzykowska, K., Eds.; Springer: Cham, Switzerland, 2020; Volume 1032, pp. 247–256. [CrossRef]
24. Klimczak, T.; Paś, J. Reliability and operating analysis of transmission of alarm signals of distributed fire signaling system. *J. KONBiN* **2019**, *49*, 165–174. [CrossRef]
25. Bednarek, M.; Dąbrowski, T.; Olchowik, W. Selected practical aspects of communication diagnosis in the industrial network. *J. KONBiN* **2019**, *49*, 383–404. [CrossRef]
26. Paś, J.; Rosiński, A.; Białek, K. A reliability-exploitation analysis of a static converter taking into account electromagnetic interference. *Transport Telecommun.* **2021**, *22*, 217–229. [CrossRef]
27. Paś, J.; Jakubowski, K. Indicator Analysis of Security Risk for Electronic Systems Used to Protect Field Command Posts of Army Groupings. *J. KONBiN* **2020**, *50*, 43–61. [CrossRef]
28. Chmielińska, J.; Kuchta, M.; Kubacki, R.; Dras, M.; Wierny, K. Selected methods of electronic equipment protection against electromagnetic weapon. *Przegląd Elektrotechniczny* **2016**, *92*, 1–8. [CrossRef]
29. Stypułkowski, K.; Gołda, P.; Lewczuk, K.; Tomaszewska, J. Monitoring System for Railway Infrastructure Elements Based on Thermal Imaging Analysis. *Sensors* **2021**, *21*, 3819. [CrossRef]
30. Polish-European Standard. PN-EN 50131-1:2009. *Alarm Systems—Intrusion and Hold-up Systems—Part. 1: System Requirements*; Polish Committee for Standardization: Warsaw, Poland, 2009.
31. Xu, S.; Li, X. Analysis on thermal reliability of key electronic components on PCB board. In Proceedings of the 2011 International Conference on Quality, Reliability, Risk, Maintenance, and Safety Engineering, Xi'an, China, 17–19 June 2011; Huang, H.-Z., Zuo, M.J., Jia, X., Liu, Y., Eds.; IEEE: Xi'an, China, 2011; pp. 52–54. [CrossRef]
32. Milic, M.; Ljubenovic, M. Arduino-Based Non-Contact System for Thermal-Imaging of Electronic Circuits. In Proceedings of the 2018 Zooming Innovation in Consumer Technologies Conference (ZINC), Novi Sad, Serbia, 30–31 May 2018; pp. 62–67. [CrossRef]
33. Shilo, G.; Ogrenich, E.; Kulyaba-Kharitonova, T.; Buhaiev, O. Thermal design of the Electronic Equipment Enclosures with Natural Air Cooling. In Proceedings of the 9th International Conference on Advanced Computer Information Technologies (ACIT), Ceske Budejovice, Czech Republic, 5–7 June 2019; pp. 153–156. [CrossRef]
34. Qiang, G.; Ya, Z.; Jinhua, Z. Dynamic Reliability Testing about Temperature Characteristic of Components. In Proceedings of the 2009 International Conference on Wireless Networks and Information Systems, Shanghai, China, 28–29 December 2009; Luo, Q., Tan, H., Eds.; IEEE: Los Alamitos, CA, USA, 2009; pp. 257–258. [CrossRef]
35. Vasile, D.C.; Svasta, P.M. Temperature sensitive active tamper detection circuit. In Proceedings of the 2017 IEEE 23rd International Symposium for Design and Technology in Electronic Packaging (SIITME), Constanta, Romania, 26–29 October 2017; IEEE: Piscataway Township, NJ, USA, 2017; pp. 175–178. [CrossRef]
36. Vasile, D.-C.; Svasta, P.; Pantazică, M. Preventing the Temperature Side Channel Attacks on Security Circuits. In Proceedings of the 2019 IEEE 25th International Symposium for Design and Technology in Electronic Packaging (SIITME), Cluj-Napoca, Romania, 23–26 October 2019; IEEE: Piscataway Township, NJ, USA, 2019; pp. 244–247. [CrossRef]
37. Wang, X.; Liu, X.; Ding, Y.; Hang, C.; Wu, G.; Liu, W.; Li, J. Study on the Low Temperature Reliability of Leaded Solder. In Proceedings of the 2020 21st International Conference on Electronic Packaging Technology (ICEPT), Guangzhou, China, 12–15 August 2020; IEEE: Piscataway Township, NJ, USA, 2020; pp. 1–5. [CrossRef]
38. Kościelski, M.; Sitek, J. Influence of soldering condition on structure and reliability of solder joints made in Package-on-Package technology. In Proceedings of the 2016 6th Electronic System-Integration Technology Conference (ESTC), Grenoble, France, 13–15 September 2016; IEEE: Piscataway Township, NJ, USA, 2016; pp. 1–4. [CrossRef]
39. Seehase, D.; Novikov, A.; Nowottnick, M. Resistance development on embedded heating layers during climatic test. In Proceedings of the 2017 21st European Microelectronics and Packaging Conference (EMPC) & Exhibition, Warsaw, Poland, 10–13 September 2017; Dziedzic, A., Jasiński, P., Eds.; IEEE: Neumarkt-St. Veit, Germany, 2017; pp. 1–5. [CrossRef]
40. Kofanov, Y.N.; Sotnikova, S.Y.; Subbotin, S.A. Method of increasing the reliability of on-board electronic equipment with an analysis of reserves for the electrical, thermal and mechanical loads. In Proceedings of the 2016 IEEE Conference on Quality Management, Transport and Information Security, Information Technologies (IT&MQ&IS), Nalchik, Russia, 4–11 October 2016; Shaposhnikov, S., Ed.; IEEE Russia North West Section: St. Petersburg, Russia, 2016; pp. 94–98. [CrossRef]
41. Stawowy, M.; Olchowik, W.; Rosiński, A.; Dąbrowski, T. The Analysis and Modelling of the Quality of Information Acquired from Weather Station Sensors. *Remote Sens.* **2021**, *13*, 693. [CrossRef]

42. Duer, S.; Duer, R.; Mazuru, S. Determination of the expert knowledge base on the basis of a functional and diagnostic analysis of a technical object. *Nonconv. Technol. Rev.* **2016**, *20*, 23–29. Available online: http://revtn.ro/index.php/revtn/article/view/115/76 (accessed on 10 July 2021).
43. Duer, S. Assessment of the Operation Process of Wind Power Plant's Equipment with the Use of an Artificial Neural Network. *Energies* **2020**, *13*, 2437. [CrossRef]
44. Burdzik, R.; Konieczny, Ł.; Figlus, T. Concept of on-board comfort vibration monitoring system for vehicles. In Proceedings of the Communications in Computer and Information Science, Activities of Transport Telematics 13th International Conference on Transport Systems Telematics, TST 2013, Katowice-Ustroń, Poland, 23–26 October 2013; Mikulski, J., Ed.; Springer: Berlin/Heidelberg, Germany, 2013; Volume 395, pp. 418–425. [CrossRef]
45. Kostrzewski, M. Analysis of selected acceleration signals measurements obtained during supervised service conditions—Study of hitherto approach. *J. Vibroeng.* **2018**, *20*, 1850–1866. [CrossRef]
46. Kukulski, J.; Jacyna, M.; Gołębiowski, P. Finite Element Method in Assessing Strength Properties of a Railway Surface and Its Elements. *Symmetry* **2019**, *11*, 1014. [CrossRef]
47. Paś, J.; Siergiejczyk, M. Interference impact on the electronic safety system with a parallel structure. *Diagnostyka* **2016**, *17*, 49–55. Available online: http://www.diagnostyka.net.pl/pdf-62677-17828?filename=Interference%20impact%20on.pdf (accessed on 10 July 2021).
48. Suproniuk, M.; Paś, J. Analysis of electrical energy consumption in a public utility buildings. *Przegląd Elektrotech.* **2019**, *95*, 97–100. [CrossRef]
49. Łukasiak, J.; Rosiński, A. Analysis of exploitation process in the aspect of readiness of electronic protection systems. *Diagnostyka* **2017**, *18*, 37–42. Available online: http://www.diagnostyka.net.pl/pdf-79784-17618?filename=Analysis%20of%20exploitation.pdf (accessed on 10 July 2021).
50. Military Handbook. *Reliability/Design Thermal Applications*; MIL-HDBK-251; Department od Defence: Washington, DC, USA, 1978.
51. Ćwirko, J.; Ćwirko, R. Temperature testing of electronic modules. *Biul. WAT* **2008**, *LVII*, 133–142.
52. Polish-European Standard. PN-EN 50130-5:2012. *Alarm Systems—Part. 5: Environmental Test Methods*; Polish Committee for Standardization: Warsaw, Poland, 2012.
53. Defense Standard. NO-04-A004-1:2016. *Military Installations—Alarm Systems—Part 1: General Requirements*; Military Centre for Standardization, Quality and Codification: Warsaw, Poland, 2016.
54. Polish-European Standard. PN-EN IEC 60721-3-3:2019-10. *Classification of Environmental Conditions—Part 3-3: Classification of Groups of Environmental Parameters and Their Severities—Stationary Use at Weatherprotected Locations*; Polish Committee for Standardization: Warsaw, Poland, 2019.
55. Polish-European Standard. PN-EN IEC 60721-3-4:2019-10. *Classification of Environmental Conditions—Part 3-4: Classification of Groups of Environmental Parameters and Their Severities—Stationary Use at Non-Weatherprotected Locations*; Polish Committee for Standardization: Warsaw, Poland, 2019.

 *energies*

Article

Operational Analysis of Fire Alarm Systems with a Focused, Dispersed and Mixed Structure in Critical Infrastructure Buildings

Krzysztof Jakubowski [1], Jacek Paś [2,*], Stanisław Duer [3] and Jarosław Bugaj [4]

[1] National Cyber Security Centre, Gen. Buka 1, 05-119 Legionowo, Poland; krzysztof.jakubowski.wel@gmail.com
[2] Division of Electronic Systems Exploitations, Faculty of Electronics, Institute of Electronic Systems, Military University of Technology, 2 Gen. S. Kaliski St., 00-908 Warsaw, Poland
[3] Department of Energy, Faculty of Mechanical Engineering, Technical University of Koszalin, 15-17 Raclawicka St., 75-620 Koszalin, Poland; stanislaw.duer@tu.koszalin.pl
[4] Division of Radiocommunication, Faculty of Electronics, Institute of Communications Systems, Military University of Technology, 2 Gen. S. Kaliski St., 00-908 Warsaw, Poland; jaroslaw.bugaj@wat.edu.pl
* Correspondence: jacek.pas@wat.edu.pl

Abstract: The article presents issues regarding the impact of operating conditions on the functional reliability of representative fire alarm systems (FASs) in selected critical infrastructure buildings (CIB). FAS should operate correctly under variable environmental conditions. FASs ensure the safety of people and CIB. Operational measurements for 10 representative systems were conducted in order to determine the impact of environmental conditions on FAS reliability. Selected operational indices were also determined. The next stage involved developing two models of representative FASs and the availability, pre-ageing time and operating process security indices. Determining operational indices is a rational selection of FAS technical and organizational solutions that enables the reliability level to be increased. Identifying the course of the FAS operating process security hazard changes in individual system lines, particularly at the initial operation stage, enables people that supervise the operation to affect operating parameters on an ongoing basis. The article is structured in the following order: issue analysis, FAS power supply in CIB, operational test results, selected FAS operating process models, determination of operational and security indices, and conclusions.

Keywords: fire alarm system; critical infrastructure; reliability requirements; construction facilities

Citation: Jakubowski, K.; Paś, J.; Duer, S.; Bugaj, J. Operational Analysis of Fire Alarm Systems with a Focused, Dispersed and Mixed Structure in Critical Infrastructure Buildings. *Energies* **2021**, *14*, 7893. https://doi.org/10.3390/en14237893

Academic Editor: Mohamed Benbouzid

Received: 7 November 2021
Accepted: 18 November 2021
Published: 25 November 2021

Publisher's Note: MDPI stays neutral with regard to jurisdictional claims in published maps and institutional affiliations.

1. Introduction

Fire alarm system (FAS) devices are security devices and, within the meaning of the Regulation of the European Parliament and of the Council No. 305 of 9 March 2011 (CPR), are treated and marketed in specific countries as building materials [1,2]. This is why all FAS devices and elements have been recognized as building products that are permanently built-in within a given building, such as floor beam, doors, windows, stairs, lintels or other building materials [3]. Owing to their function, namely protection of life, environment or accumulated movable and immovable property in buildings supervised by FAS, all devices, accumulator banks constituting backup power sources, FAS elements and modules are very important in terms of the fire safety of a given building or structure [4,5] (Figure 1).

The need to use a fire alarm system in a given building or structure may arise from the following provisions or assumptions:

- Legislation in force within a given country (e.g., in Poland, Regulation of the Minister of Interior and Administration of 7 June 2010, (Dz.U. 109, item 719) [1];
- Recommendations of competent fire safety (State Fire Service—PSP) or environmental protection authorities (e.g., District Building Supervision Office) [6,7];

37

- FAS installation and operation follows a statement or independent decision of an investor, buyer or tenant of a given building; the owner, administrator or user of a given building shall be responsible for operating the system [2].

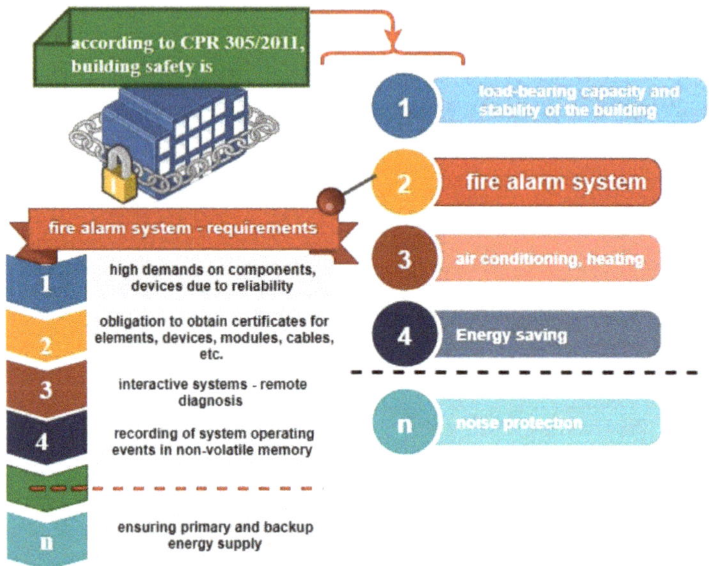

Figure 1. Requirements in the field of FAS application in buildings pursuant to CPR 305/2011.

In general, all FASs, due to their design, supervised area expressed in, e.g., m^3, the number of fire zones, number of detection circuits and lines, area coverage of protected facilities and fire hazard categories can be divided to three various organization structures [8,9]:

- Fire alarm systems of focused structure, where all loops, B-type radial lines, control lines, audio and optical signaling devices, etc. are connected to a fire alarm control unit (FACU) [1,10]. A single so-called connection node is present in the FACU in such a case. The distance of the most remote locations monitored by the FACU does not exceed the permissible detection line or circuit length due to the alarm signal transmission process [11,12]. The alarm control unit contains a single so-called connection node [13] (Figure 2).
- Fire alarm systems of the so-called distributed and scalable structure, have an advantage of simple system expansion through adding, e.g., a control device of an alarm sub-panel to a circuit. It is achieved through hooking-up additional fire alarm system sub-panels, which are slaves to a master FACU [14,15]. In such a case, A-type detection circuits, B-type radial lines with detectors and audio–optical signaling devices monitor separate fire zones within buildings that may be located over a vast area. In this case, the power and backup supplies are routed to each fire alarm control unit separately, from another internal supply line (ISL) [16]. All of these measures are introduced in order to ensure reliability and due to the current power load within the facility [17,18]. All FACUs monitoring a given structure or such a vast area in terms of fire safety are connected through a double transmission cable, a so-called ring, for reliability purposes or operate in a so-called star system, where a master FACU is located in its central place [19,20]. In the case of these systems, the distance of remote locations supervised by a given FACU can exceed the permissible length of lines, detection circuits or control lines [21,22]. Due to the costs of executing a given fire system (e.g., execution of several detection lines to a remote protected part or facility) over a vast

area, such a solution may be cheaper or more reliable than, e.g., the application of a focused FAS [23,24]. Due to the potential electromagnetic interference over a vast area, it is possible to connect individual FACUs using a fiber optic cable [25,26] (Figure 2).

- Fire alarm systems of mixed structure are executed taking into account the sole costs of executing a given FAS, but also due to the possibility of applying various complex reliability structures when integrating the entire system [27,28]. In such a case, two different FAS structures shall be used, one distributed for monitoring a vast area and a high number of buildings, and a focused one [29,30]. A focused-structure fire alarm system is in such a case connected to a distributed FACU via a transmission line. A focused or distributed system in a given building or facility may be additionally fitted with a fixed fire equipment (FFE) gas suppression system (GSS) [31,32] (Figure 2).

Currently, all FACUs used in large buildings are manufactured in the so-called modular system [33]. A FACU consists of numerous unified modules encased usually in standard metal casings [3,34]. Individually or combined into so-called sets (nodes), they can be placed at various locations of a protected building. These points may be significantly distant from each other [35,36]. The distances between individual sub-panels (nodes) may be up to 1200 m in the case of using a copper cable or even up to 15,000 m when using a single-mode optical fiber as the transmission bus [37,38]. Individual FAS sub-panels within a single node are also interconnected via a common, double (redundant) digital transmission bus [39,40] (Figure 2). Individual FAS systems may be supervised through, e.g., a single alarm receiving center (ARC) that can be located within such a vast area [41]. Due to the facilities located within such a vast area, which belong to the so-called critical infrastructure, there is a possibility to install individual, less-complex ARCs to monitor selected FASs [42,43]. All ARCs are intercommunicated and have two independent tele-transmission channels, hardwired and wireless, diagnosed with a fixed or variable processing period due to a short-circuit, opening or replacement of a transmitter or receiver [44,45]. Telecom connections used to transmit alarm, diagnostic or damage signals to ARC or PSP are established in order to ensure an appropriate reliability level and the information transmission quality [1,46].

The rest of the article is organized as follows: Section 2 is a critical review of the source literature on the current state of the problem. Section 3 contains basic information on FAS power supply, including the backup power solution, system protection method and fire switch location. Section 4 presents the results of operational tests involving two representative FAS operated in critical infrastructure facilities. Section 5 includes the developed models—operational graphs of selected FAS and computer simulation results with charts and reliability characteristics. The last, Section 6 contains conclusions arising from the conducted tests and computer simulations.

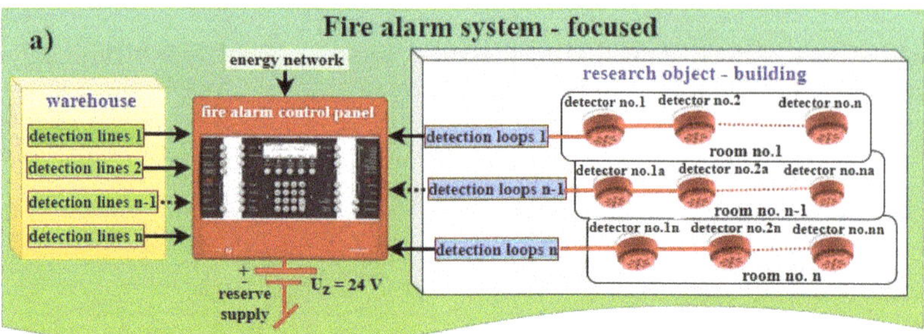

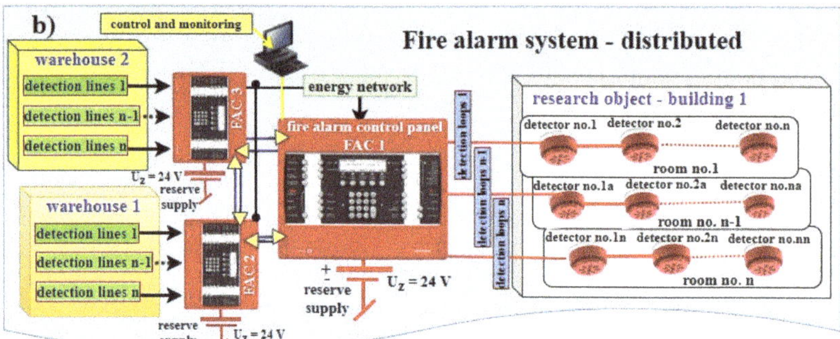

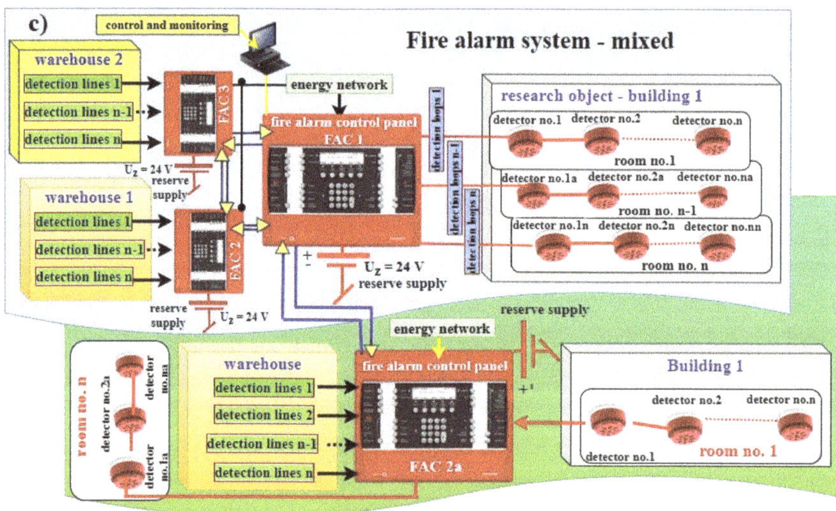

Figure 2. Configuration of fire alarm systems in buildings and structures of the so-called critical infrastructure: (**a**) simple-structure FAS, (**b**) distributed-structure FAS, (**a**,**c**) mixed-structure FAS, where U_z = 24 V—supply voltage for FAS, FAC 1, FAC 2, FAC 3, FAC 2a—power supply through fire alarm control units (FACU); building, warehouse, warehouse 1, warehouse 2—critical infrastructure building areas monitored by detection circuits and lines with fire detectors of successive addresses: 1, 2, n − 1, n; 1a, 2a, n − 1a, na, … ,1n, 2n, n − 1n, … ,nn; room No. n—monitored only by a B-type radial line as a separate fire zone, with a switching station supplying the entire vast facility of the so-called critical infrastructure.

2. Literature Review

Temperature is one of the important factors within the so-called fire triangle. The authors of [18,26,40,46] discussed temperature-based analysis covering the reliability of key electronic subsystems. This analysis enabled the optimization of, e.g., the arrangement of electronic subassemblies in the device and on the PCB, based on a conducted temperature analysis using finite element modelling (FEM) and an Ansys computer analysis. The developed method applies only to a single electronic device located at a specific place, and not electronic systems (e.g., FAS) that are spatially arranged in different rooms. Temperature is an important environmental factor determining FAS reliability. The FAS operational test conducted by the authors enables the impact of this parameter on reliability to be assessed through defining, e.g., damage intensities.

A significant issue when determining the impact of environmental factors on FAS operation is taking into account an appropriate fire detector selection [4,11,28]. In their work on fire detectors, the authors described dynamic tests involving the impact of fire characteristic values on the very detection of this phenomenon, as well as time. However, no reliability tests concerning, e.g., temperature and humidity properties for electronic subassemblies or the entire FAS were conducted. Through the available FAS damage statistics and the environmental conditions pertaining to the rooms, the authors were able to determine detector or element damage intensities.

Additional electronic circuits located in the detectors and alarm control panels, intended for diagnosing their technical condition, are very important with regard to FAS [3,9,15,46]. The articles discuss general proposals regarding measurement systems for conducting dynamic measurements of the technical conditions using various diagnostic techniques. However, the studies did not take into account the impact of signal interference arising from the application of long circuits or transmission or detection lines that are not hooked-up to an alarm control panel. In this article, the authors mitigated such errors through conducting actual tests, observing alarm, damage and detection signals in control panels, where the waveform conditioning and development process takes place.

An important issue that should be taken into account within the FAS operating process are electromagnetic interference [32,43], maximum permissible temperature changes in rooms with installed detectors [47,48] or accelerations of vibrations and shocks in electronic elements [10,49]. Within the studies discussed in the aforementioned articles, individual factors disturbing the FAS operating process are considered separately. The authors of the article conducted actual FAS tests, where the aforementioned interference was taken into account in detection and alarm signals processed in the alarm central panel. In the course of the studies conducted by the authors of the article, it was possible to physically verify, e.g., a given damage or alarm signal types in a given room, generated by interference factors.

The authors of [15,17,50] describe quality and reliability studies involving the power supply of electronic and security systems. The work presented general technical requirements, but not studies on the operation of FAS power supply, especially at the initial fire stage or under any interference. During the FAS operational studies, the authors conducted tests associated with the actual power supply process for these systems through identifying changes in detection, alarm or damage currents, as well as changes to the load of an FAS power supply in all operating modes. For the purposes of the tests, the authors also took into account the transitions to FAS backup supply (battery banks) that were often executed during a forced power grid failure, as well as through deliberate shutdown. These studies were taken into account when calculating FAS operating indices.

The authors of [8,51,52] addressed the issues associated with the transmission of alarm and damage signals to an alarm receiving center or the fire department. These articles only took into account the impact of reliability, quality, availability or time of transmission to remote points. The notification and service response time in the event of damage adopted for the analysis conducted by the authors was taken into account in the μ recovery parameter. It is particularly important for restoring an FAS to a state of fitness, which was taken into account in the system models developed by the authors of the article.

As stated in [11,24,30,34], modern technical solutions (e.g., fuzzy sets, neural networks, multisensor, optical or laser detectors, etc.) are currently applied in order to reduce, e.g., a false alarm probability, and to increase detector operational reliability, including responses to various excitations of fire sensors. The FAS analysis and operational tests conducted by the authors took into account all currently operated modern detectors that are installed in detection lines and circuits. This was considered in FAS operational graphs and models in the form of the actual numerical values for the indicators of λ damage and μ repair intensities.

Due to the particular tasks performed in facilities, FAS must be characterized by a high hardware reliability [6,21,31,37], and also as a decision on the state of security within the monitored area that is worked out based on the data received from detectors [5,11,33,34]. In the context of the tasks described by the authors of the article, especially the second of the aforementioned aspects may be considered as a diagnostic and measurement issue, taken into account during FAS operation. For obvious reasons, the manufacturers of the analyzed FAS group do not publish information on damage, reliability and replacement time of damaged elements or the effectiveness of their solutions. The studies conducted by the authors of the article verify the application of actual FAS for the fire protection of facilities. The execution of the tests was significantly hindered due to access to FAS operated in critical infrastructure buildings. This is why the article does not state the name of the facility, manufacturer of the devices or the very organization of the FAS technical structure.

Available source literature [13,25,53,54] contains various technical and operational solutions that should be taken into account when minimizing a fire in a facility or already at the evacuation stage. The authors did not address these issues; however, the FAS operational analysis itself or the test results, e.g., availability factor, can be used by designers and users of such systems to select a specific FAS manufacturer or structure. Such an analysis was not conducted during conversations with company representatives, users or the service.

3. Power Supply Implementation for Fire Alarm Systems

Proper FAS power supply organization is a very important issue, and for distributed systems in particular. Section 3 presents general requirements and proper organization of a power supply system, including the location of backup power sources, power protections and the main fire switch. Figure 3 also shows the flows of diagnostic signals with information on FAS technical conditions and power supply. Special process and technical solutions, including, e.g., appropriate positioning of the main fire switch, are used in order to ensure power supply continuity for FAS operated in the case of a fire [52,55] (Figure 3). So-called internal circuits supplying individual FACUs are executed in a building power system upstream of a so-called main switch, which deenergizes a given structure in the case of a fire event [54,55]. The FASs utilized primary mains supply and so-called backup power supply to ensure power supply continuity [56,57] (Figure 3).

Due to the power demand and rated currents flowing in the lines and circuits, the FAS power supply voltage is 24 V, unlike other security systems also operated in such facilities, e.g., closed-circuit TV (CCTV), access control system (ACS) or the intrusion detection system (IDS) [50,51]. Such a solution enables individual FASs to be supplied after isolating power from the power grid, the so-called primary power [49,50]. This allows the FAS to function without drawing power from the battery bank, the capacities of which are determined in the form of developing a so-called energy balance, i.e., electricity demand, taking into account alarm and detection currents for all FAS elements and devices [58,59]. In the case of distributed FAS, each subsystem has an individually determined energy balance [57,60]. In the case of currently operated FAS, there are no legal provisions, regulations and adequate guidelines on the operational reliability of lines, elements, devices or entire technical facilities, especially within critical infrastructure [53,59].

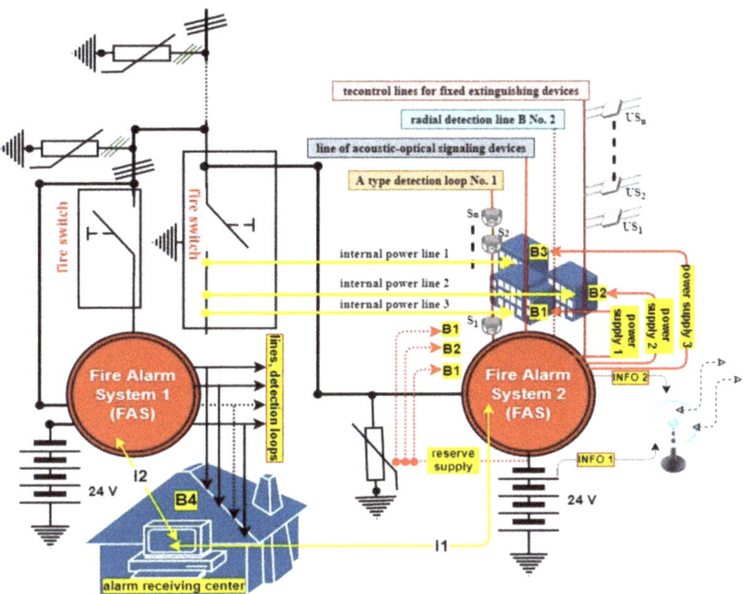

Figure 3. Power supply of fire alarm systems for B1,B2,B3—focused power supply provided by FAS2 power supply device and a 24 V battery bank for B4 building, B1,B2,B3 buildings of the so-called critical infrastructure, located over a vast monitored area are supplied from a single power line in the distributed mode, where S1,S2, … ,Sn fire detectors on A-type detection line No. 1, US1, US2, … , USn—control devices on detection line, B1,B2,B3,B4—critical infrastructure buildings, INFO1, INFO1—information (diagnostic signal) to alarm and damage signal transmission device (ADSTD) on the technical condition of primary and backup power supply systems, I1,I2—information on the technical condition of primary and backup power supply systems for the alarm receiving center, lines: A—detection circuit, B—radial line connected to a fire system control unit.

4. Determination of Operating Process Indicators for Selected Fire Alarm Systems

Determining FAS reliability is based on conducting limited tests of actual systems. Access to tests involving these systems is limited and difficult due to the location of FAS in critical infrastructure facilities. The manufacturers of individual FAS devices do not disclose reliability indices in their catalogues. This is why developing complete statistics regarding the occurrence of damage and the entire recovery process is an important issue. Such data is not published by leading FAS manufacturers. Operational tests covered all FAS components. The results were used as a basis to develop graphs that enable the min, max and mean to be read, as well as standard deviation for repair times. Graph 6 shows the most common damage types for all 10 conducted FAS tests. Studies aimed at determining the basic operating process indicator, e.g., reliability of different FAS structures, requires data to be obtained on the operation of these systems under various environmental conditions [2,52] (Figure 4). Environmental conditions (e.g., temperature, humidity, pressure, etc.) significantly impact such issues as damage intensity λ of individual FAS elements, modules or devices [11,54]. Damage λ intensity is also affected by variable supply voltage parameters, conducted and radiated electromagnetic interference, surges and voltage decays or dips [34,55]. Ten different FASs operated in various environments were studied to calculate the intensity indices for damage λ and recovery μ. Elements, modules and devices within these systems are located indoors and are exposed to direct action of a variable Earth's environment [2]. The FAS operating process analysis was conducted based on event log data recorded in the FACU. All FASs were operated within a single country, which is why it was assumed that they worked under similar environmental conditions.

Over 80,000 entries on operating events for various manufacturers of these systems were identified in FACU permanent memories [22]. Based on FACU data and face-to-face conversations with responsible persons supervising the operating process, it can be concluded that damage most usually occurred due to an operator error—so-called human factors, such as mechanical factors—inadvertent line/circuit damage during building renovation, change of environmental conditions under which the FAS is operated, external factors not attributable to the operator, e.g., voltage decay or dip and surges in FAS power supply lines, including lightning discharges that may damage vulnerable FAS electronic elements, the implementation of an incorrect FAS design, a system that is not consistent with the recommendations of the manufacturer or the operation of individual FAS subassemblies, devices or elements [7,22].

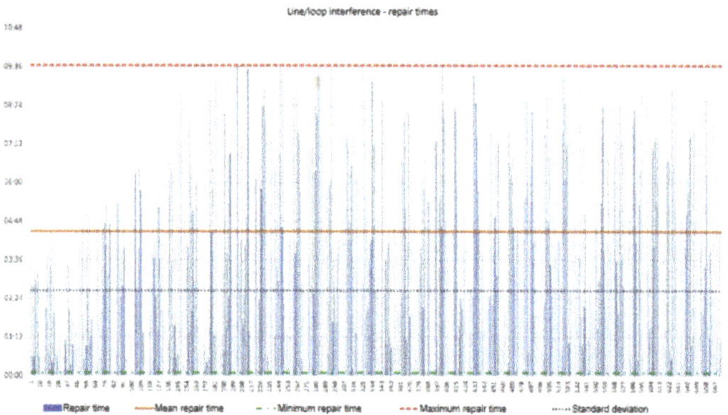

Figure 4. Repair times: line/circuit interference damage. Test results: mean repair time 268 min, minimum repair time 5 min, maximum repair time 578 min, repair time standard deviation 158 min.

Minimum, maximum and mean repair times for individual FAS elements were calculated, among others, based on the entries in the FACU event log. Standard deviation was also determined for individual repair times. Figures 4 and 5 show only selected graphs broken down by failure type identified within an FAS, while Figure 6 shows collective data for 10 FAS.

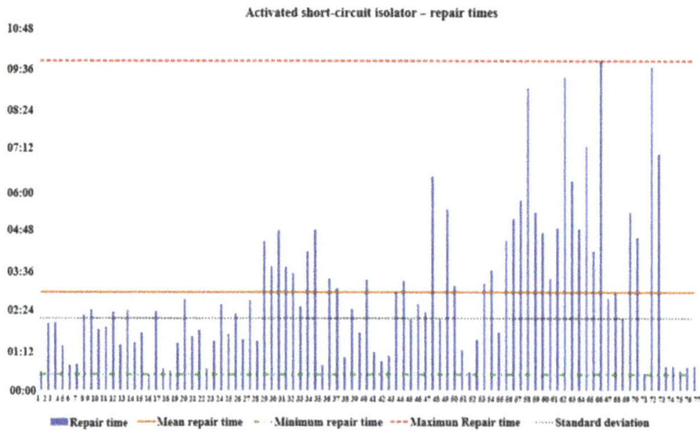

Figure 5. Repair times: activated short-circuit isolator. Test results: mean repair time 176 min, minimum repair time 29 min, maximum repair time 590 min, repair time standard deviation 129 min.

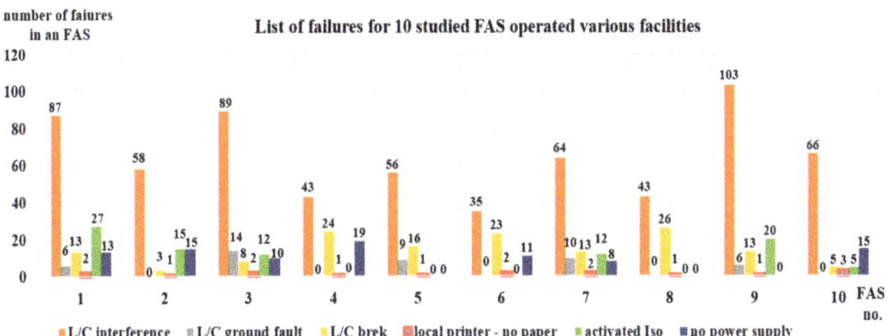

Figure 6. List of failures for 10 studied FAS operated in various facilities, where L—detection line, C—detection circuit, Iso—short-circuit isolator.

5. Operation Process of Selected Fire Alarm Systems

Section 5 contains developed operating graphs for two representative FASs. Actual operating data set out in Section 5 enable determining reliability indicators. The basic reliability distributions and the security unreliability function for individual FAS branches were determined in this section. Such an approach to the issue of FAS operation enables the so-called tuning at the initial stage of FAS use to be determined and allows one to assess the impact of individual lines or circuits on the total fitness of the system. The article demonstrates a limited yet representative number of FASs that are operated in critical infrastructure facilities. For the sake of operational safety and the security of information forwarded to the outside, such facilities most often operate a focused, and not a distributed FAS. A focused FAS structure is usually the case in critical infrastructure facilities (storages, shelters, etc.) where flammable materials are stored (Figure 2a), e.g., two fire detectors on a detection line are connected to the FAS FACU (Figure 7).

Figure 7. Simplified FAS diagram. CSP (FACU)—fire alarm control unit, Cz1, Cz2—detectors within the system detection line.

FAS, as seen in Figure 7, can stay in eight distinguished operating states:

- S1—all system elements operate correctly—FACU, Cz1, Cz2;
- S2—the only damaged fire detector is No. 1—Cz1;
- S3—the only damaged fire detector is No. 2—Cz2;
- S4—only the fire alarm control unit—CSP is damaged;
- S5—both fire detectors damaged—Cz1, Cz2;
- S6—the only working fire detector is No. 1—Cz1;
- S7—the only working fire detector is No. 2—Cz2;
- S8—all system elements damaged—FACU, Cz1, Cz2;

A Markov chain enables a graphic visualization of all states that an FAS can remain in. This includes the probabilities of staying in a given operating state or a transition between successive states. Given the initial assumptions (FAS—fit—S1 state) and all potential FAS states, it is possible to develop a Markov chain for the operating process of this system, as shown in Figure 8 [2,35]. As can be seen in this graph, an FAS can stay in the following states:

- Full fitness S_{PZ}—if occurring in state S1;
- Safety unreliability Q_B—if occurring in state S2 or S3;
- Safety hazard Q_{ZB}—if occurring in state S4 or … S5, S6, S7, S8.

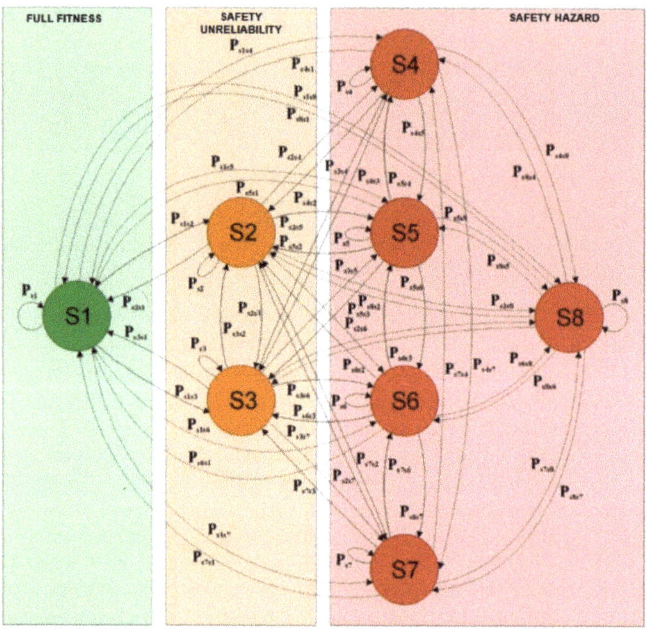

Figure 8. The Markov chain for an FAS consisting of a single FACU and two fire detectors—Cz1, Cz2, where P_{S1S2} is the probability of transition between state S1 and S2, P_{S2S1} is the probability of transition between state S2 and S1, etc., P_{s1}, P_{s2}, … , etc. are the probabilities of the FAS remaining in the distinguished states S1, S2, … , etc.

The transitions between successive FAS states are described as probability functions, e.g., probability of a system's transition from state S_x to state S_y, marked in Figure 8 as $P_{S_xS_y}$. By using the aforementioned damage and repair probability functions, it is possible to determine all possible transitions. Due to the fact that damage to FAS elements is independent of each other, through the application of design and organizational solutions, the probability of transition P_{S1S2} can be expressed by a product of the probability of damage to fire detector No. 1 (Cz1), and the probability of correct operation of the FACU and fire detector No. 2 (Cz2); this is determined by Equation (1).

$$P_{S1S2}(t) = R_{CSP}(t) \cdot Q_{Cz1}(t) \cdot R_{Cz2}(t) \tag{1}$$

By proceeding analogously in the case for other probabilities of leaving the S1 state, it is possible to determine and calculate individual probabilities occurring in the case of a focused, simple FAS, i.e., determine other P_{S1S3}, P_{S1S4}, P_{S1S5}, P_{S1S6}, P_{S1S7}, P_{S1S8}.

By adopting the operating data obtained in the course of studying 10 various FAS designs, it is possible to determine the intensities of damage λ and repairs μ for various components of such a technical structure, operated for a selected period of time. The damage λ and recovery μ intensities were determined based on studies and observations of the FAS operating process. They respectively amount to, for a focused FAS-CSP (FACU), Cz1, Cz2, which is determined by Expressions (2)–(4).

$$\lambda_{CSP} = 1.25478 \cdot 10^{-7} \left(\frac{1}{h} \right); \; \mu_{CSP} = 0.1306 \left(\frac{1}{h} \right) \tag{2}$$

$$\lambda_{Cz1} = 4.48762 \cdot 10^{-6} \left(\frac{1}{h} \right); \ \mu_{Cz1} = 0.1818 \left(\frac{1}{h} \right) \tag{3}$$

$$\lambda_{Cz2} = 3.14687 \cdot 10^{-6} \left(\frac{1}{h} \right); \ \mu_{Cz2} = 0.1810 \left(\frac{1}{h} \right) \tag{4}$$

where λ_{CSP}—FACU damage intensity, μ_{CSP}—FACU recovery intensity, λ_{Cz1}Cz1 damage intensity, μ_{Cz1}—Cz1 recovery intensity, λ_{Cz2} —Cz2 damage intensity, μ_{Cz2}Cz2 recovery intensity.

Figures 9–11 show selected P probabilities as a function of the time of transition from the S1 state to other states, and FAS residence times in various states, for t = 8760 h.

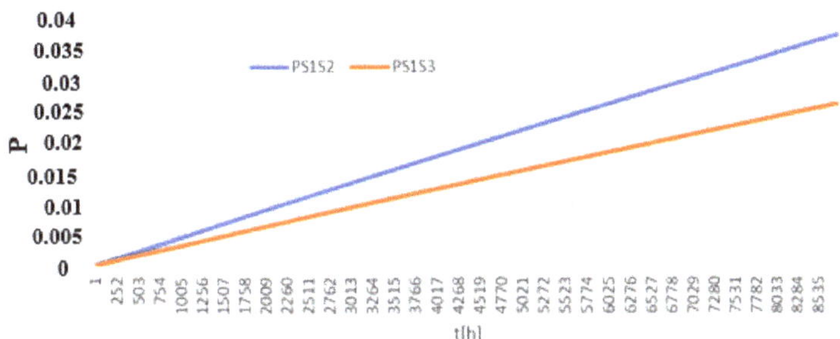

Figure 9. Probability of FAS transitioning from state S1 to states S2 and S3.

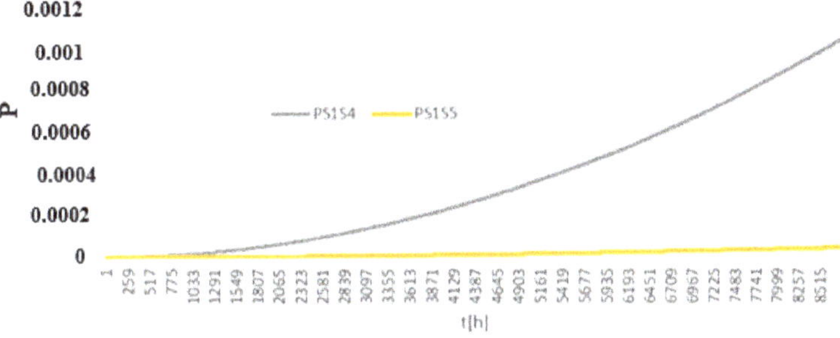

Figure 10. Probability of FAS transitioning from state S1 to states S4 and S5.

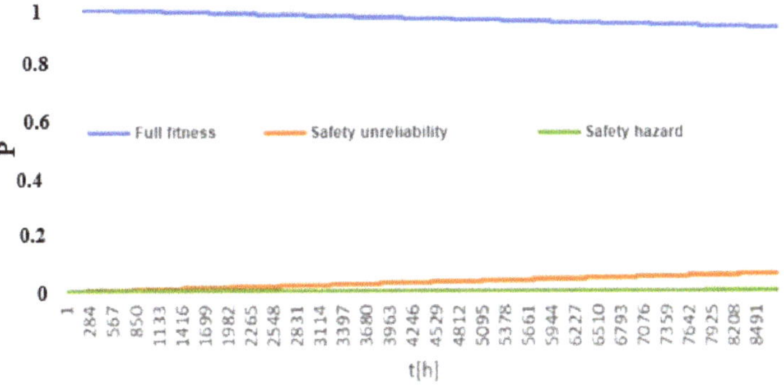

Figure 11. Probability functions for a system staying in one of the three primary states.

Based on the obtained results, it is possible to calculate the A availability coefficient for FAS, according to Expression (5).

$$A = \frac{T}{T + Q} \tag{5}$$

where A—FAS availability coefficient; T—mean duration of FAS staying in a state of full fitness or safety unreliability (mean probability); Q—mean duration of the system staying in the state of safety hazard (mean probability).

In the FAS in question, they respectively amounted to the following, according to Expression (6):

$$- T = 0.966768623; - Q = 0.000090143; - A = 0.999098569 \qquad (6)$$

In order to calculate the time spent in one of the three general safety states, it will be necessary to calculate the mean probability of staying in a given state. In the FAS in question, they respectively amounted to the following:

- $M_{PZ} = 0.966768623$ mean probability of staying in a state of full fitness;
- $M_{ZB} = 0.032329946$ mean probability of staying in a state of safety unreliability;
- $M_Z = 0.0000901431$ mean probability of staying in a state of safety hazard.

For the adopted focused-type FAS operation time t = 8760 h, the times of staying in individual states are respectively determined by Expression (7):

$$T_{PZ} = M_{PZ}*t = 8468.8934[h]; T_{ZB} = M_{ZB}*t = 283.21323[h]; T_Z = M_Z*t = 7.8965997[h] \qquad (7)$$

An FAS staying in a state of full fitness is most probable, while reaching a state of safety hazard, i.e., FACU damage, is less probable. The least probable state that may occur within the FAS in question is S8, i.e., damage to all system elements—the state of FACU, Cz1 and Cz2 unfitness. A specified number of detectors, manual call points (MCP) on detection lines, are used in the case of an FAS that is used to protect buildings and rooms and communication routes therein.

This is determined by the maximum area monitored by lines connected to a FACU. Figure 12 shows a simplified diagram of an FAS consisting of B-type lines with connected detectors, MCP, control modules or audio and optical devices. Various technical and organizational solutions that enable the preset fitness level to be achieved for the entire facility are used due to FAS operational reliability [54]. This is why the application of redundancy and a fail-safe principle in such systems already at the design stage leads to a situation in which a single FAS element or device failure does not result in so-called critical or catastrophic damage [20,53]. This is particularly important when developing a so-called FAS control matrix that takes into account the so-called fire scenario [10].

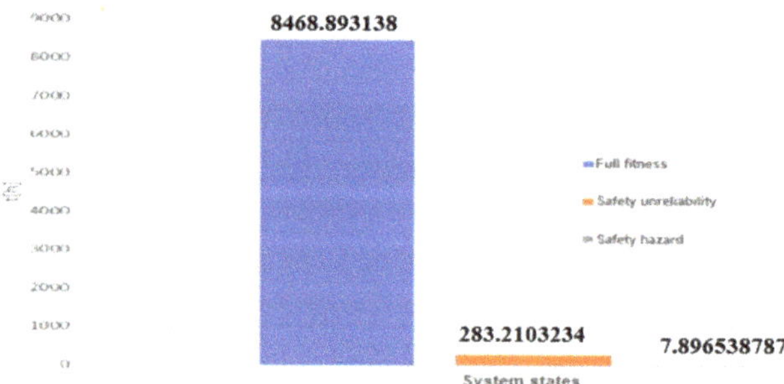

Figure 12. Focused FAS time of residence in various safety states.

The application of various solutions that take into account FAS reliability enables fire scenarios to contain operational events that involve so-called indirect system unfitness states or safety hazard states $Q_{ZB}(t)$; they are often called in source literature as efficiency failure [2,55]. There are only two basic operating states in technical systems without redundancy. These are a state of full fitness $R_o(t)$ and a state of safety unreliability $Q_B(t)$ [22,45] (Figure 8). In order to identify fire alarm system unreliability indices, it is necessary to determine the environmental conditions under which such technical structures are operated [56,57].

These environmental parameters significantly impact the damage intensity coefficient λ for elements, devices and modules making up the FAS. For the sake of fire safety within B-type detection lines (Figure 13), there are restrictions to the number of installed detectors (a maximum of 32) and MCPs (10 units) [2,45,58]. Individual fire zones within an FAS shall be separated by a short-circuit isolator (Figure 13). Figure 13 shows a focused FAS monitoring a building [22,59,60]. It consists of two separate fire zones located in the building and warehouse [2]. In order to ensure power supply security, the FAS was equipped with a backup power supply in the form of a battery bank with a capacity determined through calculating the energy balance that takes into account monitoring and alarm currents for individual elements of this system [61].

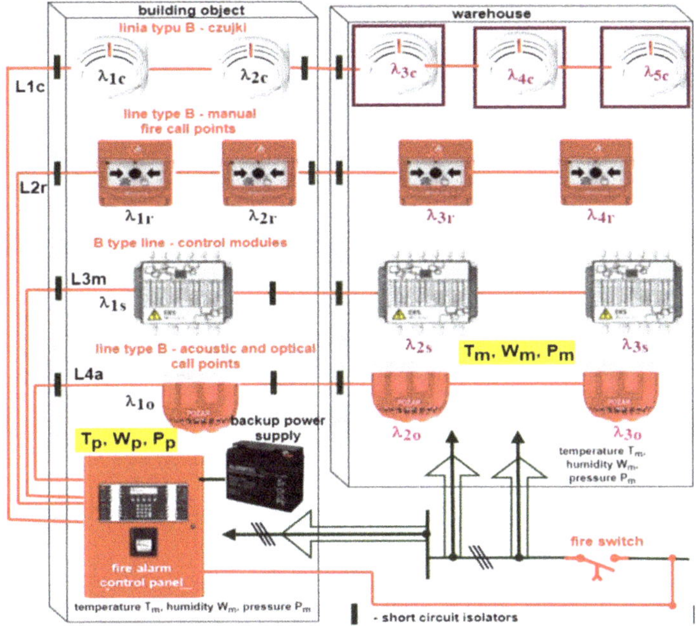

Figure 13. A focused FAS located in two different remote buildings—utility rooms, storage rooms, operation under various environmental conditions, where T_{pm}—temperature, W_{pm}—humidity, P_{pm} —pressure.

The control modules located within a B-type line enable the control of, e.g., an audio warning system (AWS) or smoke exhaust devices and dampers [62,63]. The system consists of detection circuits, some of which have programmed detectors in coincidence systems, a control loop with a module controlling fire safety devices as well as technical and safety systems in the building and the storage room [22,29].

A signaling line with audio and optical signaling devices is also hooked-up to the FACU [36]. The FAS has a serial and parallel reliability structure [2,22]. Figure 14 is a graph showing the operating process of the FAS from Figure 13.

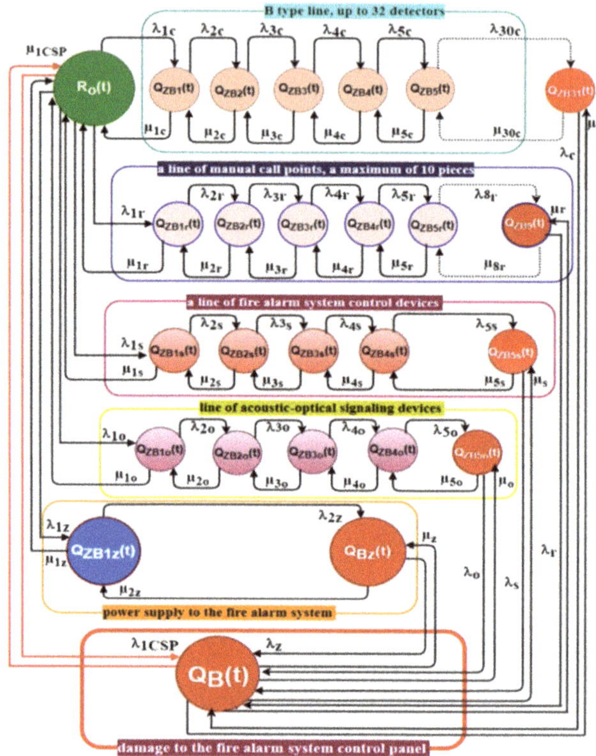

Figure 14. Operating states occurring in the course of operating a focused FAS with an addressable fire alarm control unit with detection and monitoring control on B-type radial lines, B-type audio and optical signaling device lines and MCPs, taking into account power supply unreliability—industrial mains and backup power supply.

Assuming certain baseline conditions, the focused FAS in question can be described using Expressions (8) and (9):

$$R_0(t) = 1 \tag{8}$$

$$Q_{ZB1}(0) = Q_{ZB2}(0) = Q_{ZB3}(0) = Q_{ZB4}(0) = Q_{ZB5}(0) = Q_{ZB6}(0) = Q_{ZB7}(0) = Q_{ZB8}(0) = Q_{ZB9}(0) = Q_{ZB10}(0)$$
$$= Q_{ZSA1}(0) = Q_{ZSA2}(0) = Q_B(0) = 0 \tag{9}$$

where $R_0(t)$—probability function for FAS staying in a state of full fitness S_{PZ}; $Q_{ZB1}(t)$ do $Q_{ZB31}(t)$, $Q_{ZB1r}(t)$ do $Q_{ZB9r}(t)$, $Q_{ZB1s}(t)$ do $Q_{ZB5s}(t)$, $Q_{ZB10}(t)$ do $Q_{ZB5o}(t)$, $Q_{ZB1z}(t)$—probability function for FAS staying in individual safety hazard states; $Q_B(t)$—probability function for FAS staying in the safety unreliability state S_B; $Q_{Bz}(t)$—probability function for FAS staying in the safety unreliability state S_B in the case of a power failure; $Q_{ZB5o}(t)$, $Q_{ZB5s}(t)$, $Q_{ZB9r}(t)$, Q_{ZB31}—probability function for FAS staying in the safety unreliability state SB in the case of a failure in audio and optical signaling device line, controller line, B-type line with manual call point, radial detector line, λ_{1CSP}—intensity of transition from a state of full fitness S_{PZ} to a state of safety unreliability S_B; μ_{1CSP}—intensity of transition from a state of safety unreliability SB to a state of full fitness S_{PZ}; $\lambda_{1c}, \lambda_{2c}, \lambda_{3c}, \ldots, \lambda_{1r}, \lambda_{2r}, \lambda_{3r}, \ldots, \lambda_{1s}, \lambda_{2s}, \lambda_{3s}, \ldots, \lambda_{1o}, \lambda_{2o}, \lambda_{3o}, \ldots,$ —intensity of transitions from a state of full fitness SPZ or a state of safety hazard S_{ZB} to a state of safety unreliability $Q_B(t)$, S_B—as per the designations in Figure 14; $\mu_{1c}, \mu_{1c}, \mu_{3c}, \ldots, \mu_{1r}, \mu_{1r}, \mu_{3r}, \ldots, \mu_{1s}, \mu_{1s}, \mu_{3s}, \ldots, \mu_{1o}, \mu_{1o}, \mu_{3o}, \ldots$ —intensity of transitions from a state of safety hazard S_{ZB} to a state of full fitness S_{PZ}, from a state of safety unreliability to a state of safety hazard or a state of safety hazard to a state of full fitness $R_0(t)$, as per the designations in Figure 14.

Just like in Figure 13, the probability for an FAS staying in individual safety hazard, safety unreliability and full fitness states, as well as the transition intensities and recovery values were conducted using specialized BlockSim computing software by ReliaSoft [2,58]. This software enables simulations, studying reliability and calculating system availability. It also allows one to conduct various analyses in this regard [22,64,65]. The software offers appropriate graphical interface, which enables the modelling of complex systems (also other technical structures) and processes using relevant reliability block diagrams and fault tree analysis or a combination of both aforementioned approaches. It also offers separate process flow modules and Markov diagrams that are used to conduct the simulations [7,42].

State-discrete and time-continuous Markov chains for the focused FAS, as shown in Figure 14, can be described as a graph. In such a case, the transitions between individual states are defined by (fixed) transition intensities determined for the system in question (Figure 13). It is common to use time-continuous Markov chains to analyze the issue of reliability or, e.g., system A availability [2,7]. This can be described using a set of ordinary differential equations. In such a case, each differential equation represents a change in the probability of an FAS staying in a specified state (Expression (10)) [7,22]:

$$\frac{dP_j}{dt} = \sum_{l=1}^{n} \lambda_{lj} P_l - \sum_{l=1}^{n} \lambda_{jl} P_j \tag{10}$$

where n—number of considered states of a focused FAS; P_j—probability of the considered FAS to stay in a distinguished state j; P_l—probability of the considered FAS to stay in a distinguished state l; λ_{lj}—intensity of FAS's transition from state l to state j; λ_{jl}—intensity of a FAS' transition from state j to state l.

The conditions for solving differential equations to determine FAS reliability and other security indices always result from the initial values for the probability of a system staying in each of the distinguished states. In the case of this focused FAS, the initial system states were described by Expression (3). Figure 14 shows the migration of possible states for a focused FAS with B-type open lines for detectors, MCPs and control modules, without sending alarm and damage signals to the PSP (respective graph for a focused FAS model from Figure 12)—graph printout from the BlockSim diagram [2,32].

Taking into account the determined damage λ and recovery μ intensities relative to individual elements and devices of a focused FAS shown in Figure 14, it is possible to determine specific security indices for the operating process of this system. After commissioning and first start-up, an FAS operated within a given critical infrastructure facility is in a state of fitness, $R_0(t) = 1$, which means that all elements and FACUs are efficient. At the time $t = 0$ (s), the probability of a focused FAS staying in the state $R_0(t) = 1$ (Figure 15), while the probability of staying in the state $Q_B(t)$ (safety unreliability) (Figure 16) is equal to zero. As the operation time passes, the $R_0(t)$ and $Q_B(t)$ values change. Value $R_0(t)$ decreases, as in Figure 16, at a time [0, 2400 h]; this is the beginning of the operating process.

Later on, the function waveform has a constant value equal to 0.9945, while the value of function $Q_B(t)$ grows accordingly (Figure 17). It stabilizes after a time equal to 35.04 (h), when $Q_B(t) = 3.4 \cdot 10^{-5}$. A very important issue at the beginning of the operating process is to determine the values of individual Q_{ZB} (safety hazard) functions for individual B-type detection lines hooked-up to FACU. Calculating these values will enable the impact of individual detection lines on the value of $R_0(t)$ to be determined for the entire system, or system fitness in other words.

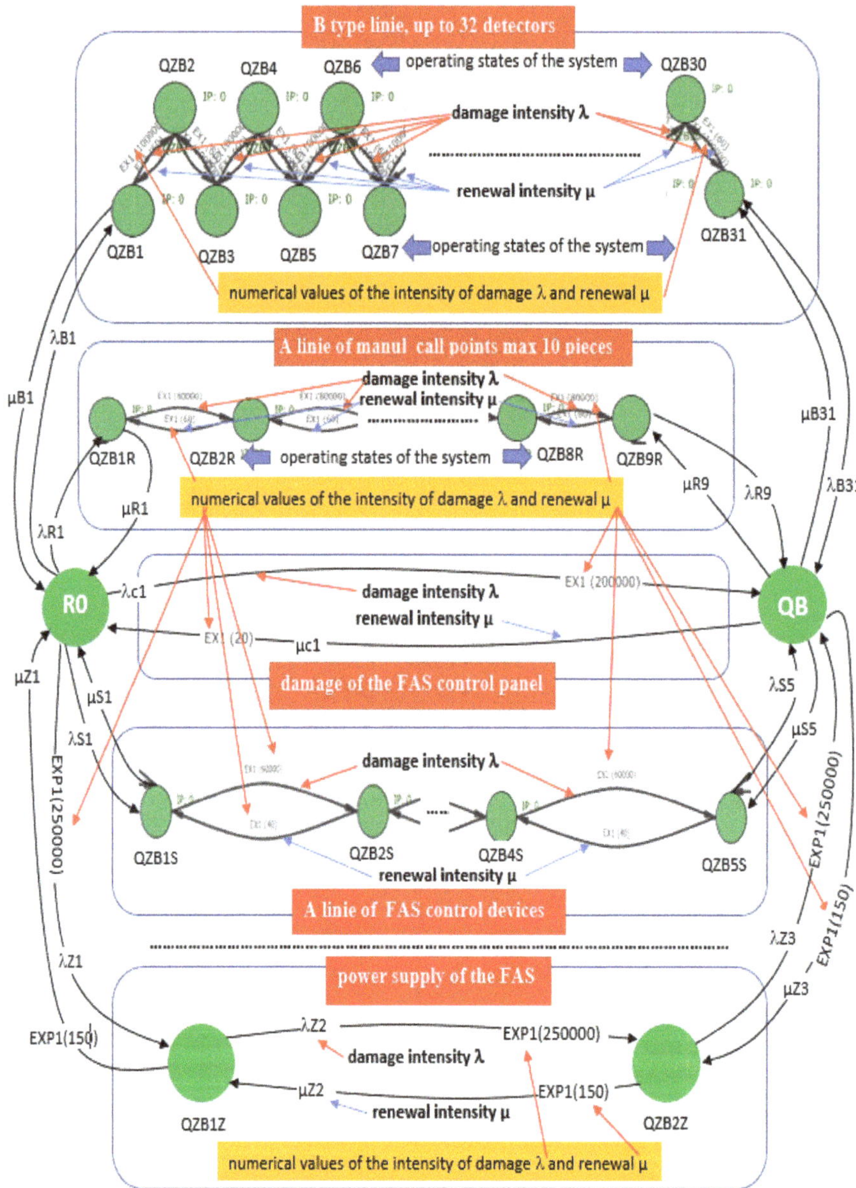

Figure 15. Migration of possible states for a focused FAS with B-type open lines for detectors, MCPs and control modules, without sending alarm and damage signals to the State Fire Service (graph printout from the BlockSim diagram).

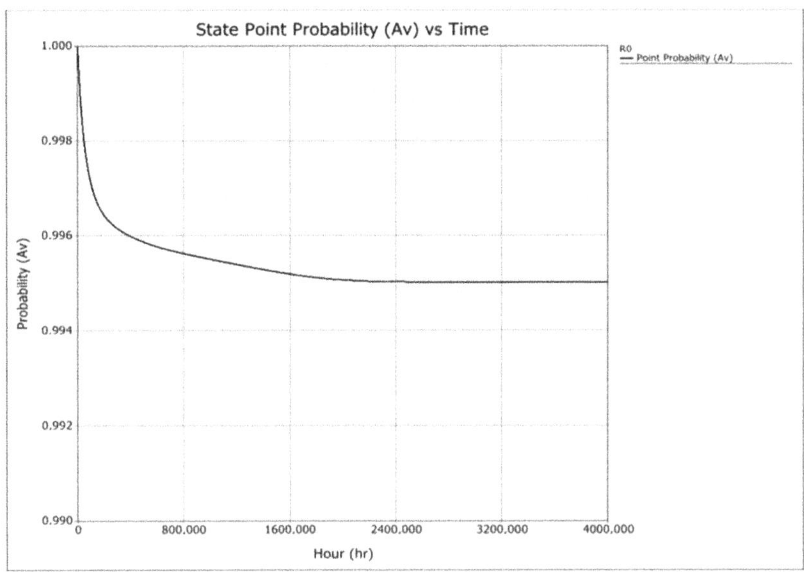

Figure 16. Probability for a focused FAS to stay in the state $R_0(t)$ with B-type open lines, without notifying the P_{SP}, as a function of time (study based on computational data from the BlockSim software).

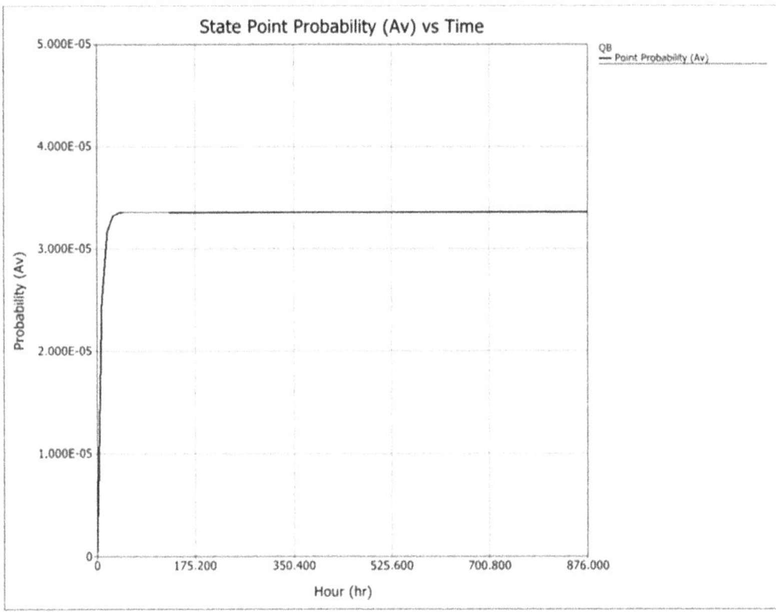

Figure 17. Probability of a focused FAS staying in the state Q_B—failure (unreliability) for the state S_0—B—type open line system, without notifying the PSP, as a function of time (study based on computational data from the BlockSim software).

After the entire FAS is activated and switches to the monitoring mode, which is the basic operation, the dominant state occurring at the initial moments of operation is state $Q_{ZB1}(t)$, which occurs within the B-type detection line. At the initial period of the FAS operation, this function increases $Q_{ZB1}(t)$—at $t = 9$ h, the value of the function is $8.35 \cdot 10^{-5}$

(Figure 17), while after 10 h of FAS operation, this value is $Q_{ZB1}(t) = 0.000162747$. Other safety hazard states occurring within other detection lines connected to FACU reach low values (Figure 18). Please note the slow increase of all safety hazard functions for individual detection lines connected to FACU at the initial operation period, i.e., up to a maximum of 180 h (Figure 18). This corresponds approximately to seven first days of the FAS operating process. This is a so-called "breaking-in process" or the "infancy", which applies to all technical, mechanical, electrical or electronic objects. The system in question is electronic, since all FAS devices contain appropriately connected and polarized electronic elements and spaces arranged on PCBs, encased in a special housing, fireproof most usually, e.g., detectors, modules, MCPs or audio and optical signaling devices. The percentage share and value of individual safety hazard states that have the potential to occur within individual detection lines connected to a FACU are presented in Figures 19 and 20a–c, respectively. Figure 21 shows a graph with the % share of individual components in a safety hazard state $Q_{ZB}(t)$, for a line with connected detectors (max 32). The highest % share among individual components for the initial operation stage, i.e., from starting the FAS up to 9 h, is achieved by component $Q_{ZB1}(t)$. It amounts to as much as 97% of the value of all components. The % share of all components $Q_{ZB1\text{-}31}(t)$ within the operating process constantly changes.

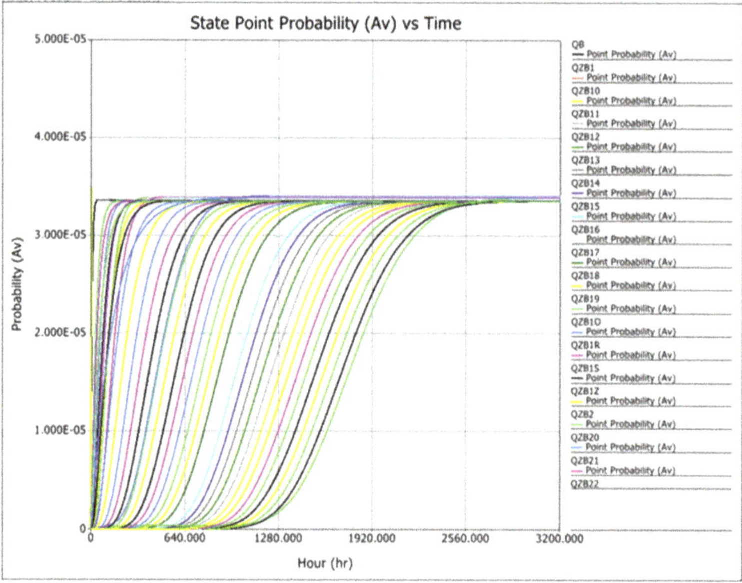

Figure 18. Probability of an FAS staying in distinguished safety hazard states $Q_{ZB}(t)$ in the case of the system in question. Zonal (partial) availability coefficients for separated technical states of a focused FAS $Q_B, Q_{ZB1}, Q_{ZB2}, Q_{ZB3}, \dots, Q_{ZB1r}, Q_{ZB2r}, Q_{ZB3r}, \dots, Q_{ZB1s}, Q_{ZB2s}, Q_{ZB3s}, \dots, Q_{ZB1o}, Q_{ZB1o}$, i.e., safety unreliability and hazard. The figure does not show state $R_0(t)$ (for $t = 0$ $R_0(t) = 1$) (study based on computational data from the BlockSim software).

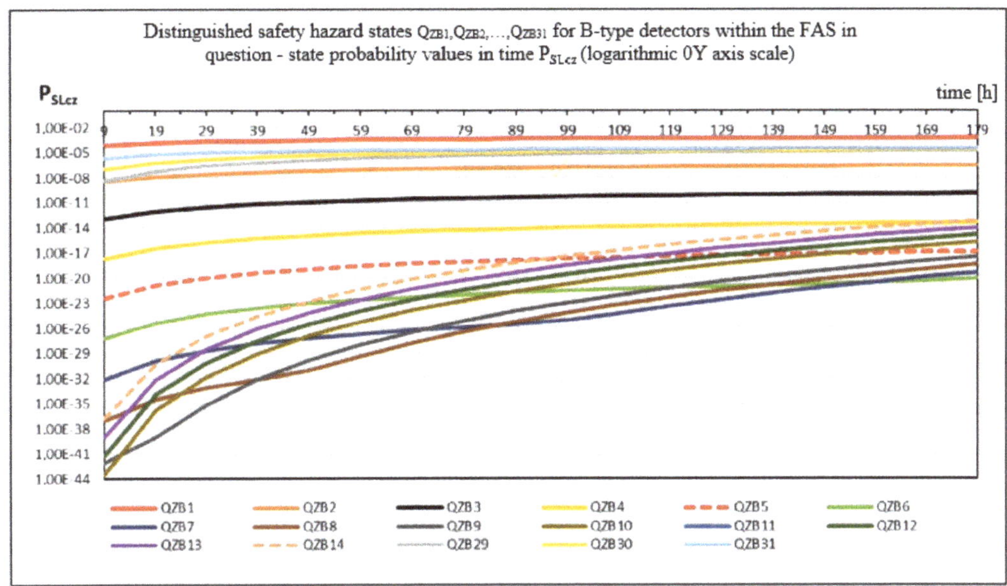

Figure 19. Zonal (partial) availability coefficients for separated technical states of a focused FAS $Q_B, Q_{ZB1}, Q_{ZB2}, Q_{ZB3},$ ... ,$Q_{ZB29}, Q_{ZB30}, Q_{ZB31}$—safety hazard indices for a B-type detector line (own study).

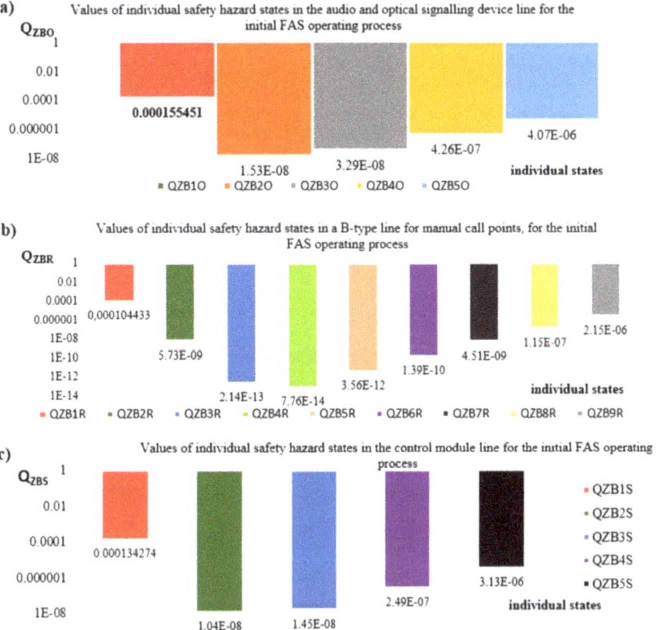

Figure 20. Values of individual safety hazard states in a focused FAS, for individual B-type lines of the system: (**a**) audio and optical signaling line, (**b**) manual call point line, (**c**) control module line (own study).

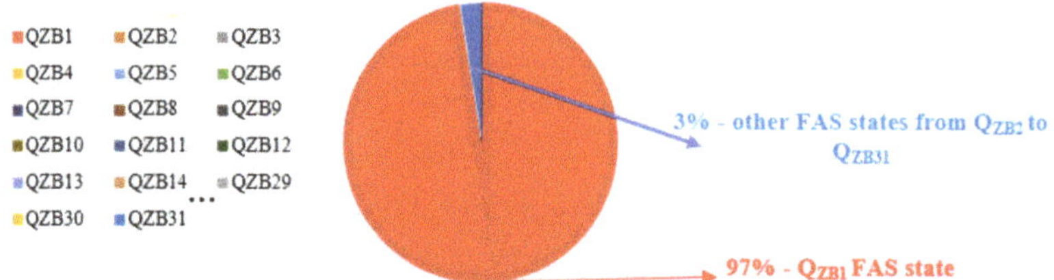

Figure 21. Percent share of individual safety hazard states Q_{ZB} of the FAS in question, at the beginning of system operation, for t = 9 h (own study).

Figure 20a–c shows the values of individual safety hazard states $Q_{ZB}(t)$ in a focused FAS, as in Figure 13, for individual B-type detection lines, which means audio and optical signaling device lines, manual call point lines and control module lines. The highest value is reached by the safety hazard component $Q_{ZB1O}(t) = 0.000155451$, i.e., a line with optical and audio signaling devices. Signaling devices are operated under varying environmental conditions, external or internal, and when the alarm is triggered, they consume significant amounts of electricity from the power supply or battery bank, constituting backup power supply sources. They consume power also when monitoring. It is used to diagnose, e.g., FACU-device connection continuity, i.e., to determine the absence of a short circuit or open-circuit of the power supply line. Individual components of the safety hazard $Q_{ZB}(t)$ reach different values (Figure 21). This difference is sometimes significant, e.g., $Q_{ZB1O}(t) = 0.000155451$, and $Q_{ZB2O}(t) = 1.53 \cdot 10^{-8}$, which means that some components are dominant within the operating process, while other are negligible at the initial operating stage. This tells a given FAS operator which detection lines and elements within such a bus should be given special attention in terms of the operating process at its initial stage. In order to better illustrate individual safety hazard components $Q_{ZB}(t)$ on detection lines connected to the FACU, the 0Y axis in Figure 20a–c has a logarithmic scale. Please note a certain regularity in the graphs in Figure 20a–c; the second or third component of safety hazard $Q_{ZB}(t)$ reaches a minimum value, e.g., $Q_{ZB2O}(t) = 1.53 \cdot 10^{-8}$, $Q_{ZB4R}(t) = 7.76 \cdot 10^{-14}$ or $Q_{ZB2S}(t) = 1.04 \cdot 10^{-8}$.

6. Conclusions

FASs are some of the most important electronic security systems operated in critical infrastructure facilities or the so-called intelligent buildings. FAS devices or elements, unlike the rest of electronic alarm systems, are governed by the Regulation of the European Parliament and of the Council (EU) No. 305/2011 of 9 March 2011 (CPR). For all FASs operated in facilities, this means that elements or devices constituting components of such systems are treated similarly to building products, e.g., a floor beams and building partitions (walls, doors, windows, etc.) that are permanently embedded into a given structure. Due to their function in a building, they are considered to be very crucial to safety, fire-related in this case. This is why it is very important to ensure operational reliability of an FAS within a given facility and to ensure a low (zero) probability of a false alarm. Appropriate technical and organizational solutions are applied in relation to these two very important issues. The article discussed two various FAS structures. Due to the scope of conducted tasks and fire controls, the reliability and operational structure of such FASs is most usually mixed.

FAS designs utilize all available organizational solutions and technical measures aimed at increasing reliability, e.g., redundancy, backups, etc. This article presents the results of operating process tests involving selected FASs and the determined intensity values for λ damage and μ recovery for selected elements and components of this system. The authors developed models of two FASs operated in critical infrastructure facilities. An operational and reliability analysis was conducted for selected system models, used later as a base to distinguish various operational states. Developed operating process models and the determined real values of λ and μ intensities enabled operational safety process indices to be determined for these systems. The A availability coefficient for a simple system consisting of a FACU and two detectors amounted to 0.999098569. This means that an FAS operated throughout the year is fit for 8752.104 h (or 99.91% of the entire year). Due to their complexity, the calculations for the second FAS, as in Figure 12, were conducted using the BlockSim simulation software by ReliaSoft. The FAS fitness probability $R_0(t)$ at the initial operating process decreases over time [0, 2400 h]; however, its further waveform has a constant value $R_0(t) = 0.9945$. The authors of the article determined the values of individual Q_{ZB} (safety hazard) functions for individual B-type detection lines hooked-up to FACU. Calculating individual Q_{ZB} functions enables determination of the impact of B-type detection lines on the value of $R_0(t)$ of the entire FAS, or system fitness in other words. Individual safety hazard functions for selected detection lines increase and stabilize after an operation time of 180 h.

This corresponds approximately to seven first days of the FAS operating process. This is a so-called "breaking-in process" or FAS "infancy", where users should pay particular attention to the operation of such an electronic system. All coefficients of the so-called zonal (partial) availability associated with safety hazard functions within the further operating process stabilize at constant levels. Quite often, the differences between individual FAS safety hazard functions are high, e.g., $Q_{ZB1O}(t) = 0.000155451$, and $Q_{ZB2O}(t) = 1.53 \cdot 10^{-8}$. For an FAS operator, this means that some components of the operating process associated with unfitness are dominant, while others, occurring within the initial operating process, are negligible. With the knowledge of such operating data, an FAS user knows which detection lines and elements within such a bus should be given special attention in terms of the operating process at its initial stage.

Author Contributions: Conceptualization, J.P., K.J. and S.D.; methodology, K.J., J.P. and S.D.; validation, J.P.; K.J. and J.B.; formal analysis, J.P., K.J. and S.D.; investigation, J.P., K.J. and S.D.; resources, J.P., K.J., S.D. and J.B.; data curation, K.J., J.P., S.D. and J.B.; writing—original draft preparation, K.J., J.P., S.D. and J.B.; writing—review and editing, J.P., K.J., S.D. and J.B.; visualization, K.J. and J.P.; supervision, J.P. and K.J.; project administration, J.P. and K.J. All authors have read and agreed to the published version of the manuscript.

Funding: This research received no external funding.

Institutional Review Board Statement: Not applicable.

Informed Consent Statement: Not applicable.

Data Availability Statement: Not applicable.

Conflicts of Interest: The authors declare no conflict of interest.

List of Important Abbreviations and Symbols

AWS	Audio Warning Systems
CPR	Council Parliament Regulation
PSP	State Fire Service
FACU	Fire Alarm Control Unit
ISL	Internal Supply Line
FFE	Fixed Fire Equipment
ARC	Alarm Receiving Centre
GSS	Gas Suppression System
CCTV	Closed-Circuit TV
ACS	Access Control System

IDS	Intrusion Detection System
ADSTD	Alarm and Damage Signal Transmission Device
λ	Intensities of Damage
μ	Intensities of Repairs
A	Availability Coefficient
MCP	Manual Call Points
$Q_{ZB}(t)$	Safety Hazard States
$R_o(t)$	State of Full Fitness
$Q_B(t)$	State of Safety Unreliability
A	Detection Circuit
B	Radial Line Connected to a Fire System Control Unit
US	Control Devices on Detection Line
P_{S1S2}	Probability of Transition Between State S_1 and S_2
P_{S1}	P_{s1} probability of the FAS remaining in the distinguished states S_1
M_{PZ}	Mean Probability of Staying in a State of Full Fitness
M_{ZB}	Mean Probability of Staying in a State of Safety Unreliability
M_Z	Mean Probability of Staying in a State of Safety Hazard

References

1. *Regulation of Ministry of the Interior and Administration of Poland (MSWiA) of 7 June 2010 (Journal of Laws of the Republic of Poland No. 109, Item 719) Concerning Fire Protection of Buildings and Other Facilities and Grounds*; Ministry of the Interior and Administration of Poland: Warsaw, Poland, 2021. Available online: https://sip.lex.pl/akty-prawne/dzu-dziennik-ustaw/ochrona-przeciwpozarowa-budynkow-innych-obiektow-budowlanych-i-terenow-17626053 (accessed on 17 November 2021).
2. Klimczak, T.; Paś, J. *Basics of Exploitation of Fire Alarm Systems in Transport Facilities*; Military University of Technology: Warsaw, Poland, 2020.
3. Duer, S. Artificial neural network in the control process of object's states basis for organization of a servicing system of a technical objects. *Neural Comput. Appl.* **2012**, *21*, 153–160. [CrossRef]
4. Keding, L. An Optimization of Intelligent Fire Alarm System for High-Rise Building Based on anasys. In *Intelligence Computation and Evolutionary Computation. Advances in Intelligent Systems and Computing*; Du, Z., Ed.; Springer: Berlin/Heidelberg, Germany, 2013. [CrossRef]
5. Jafari, M.J.; Pouyakian, M.; Khanteymoori, A.; Hanifi, S.M. Reliability evaluation of fire alarm systems using dynamic Bayesian networks and fuzzy fault tree analysis. *J. Loss Prev. Process. Ind.* **2020**, *67*, 104229. [CrossRef]
6. Pati, V.B.; Joshi, S.P.; Sowmianarayana, R.; Vedavathi, M.; Rana, R.K. Simulation of Intelligent Fire Detection and Alarm System for a Warship. *Def. Sci. J.* **1989**, *39*, 79–94. [CrossRef]
7. Grabski, F. *Semi-Markov Processes: Applications in System Reliability and Maintenance*; Elsevier: Amsterdam, The Netherlands, 2015.
8. Polak, R.; Laskowski, D.; Matyszkiel, R.; Łubkowski, P.; Konieczny, Ł.; Burdzik, R. Optimizing the Data Flow in a Network Communication Between Railway Nodes. In *Research Methods and Solutions to Current Transport Problems*; Siergiejczyk, M., Krzykowska, K., Eds.; Springer: Cham, Switzerland, 2020; pp. 351–362. [CrossRef]
9. Duer, S. Examination of the reliability of a technical object after its regeneration in a maintenance system with an artificial neural network. *Neural Comput. Appl.* **2011**, *21*, 523–534. [CrossRef]
10. Cha, J.H.; Finkelstein, M. *Point Processes for Reliability Analysis Shocks and Repairable Systems*; Springer: Berlin/Heidelberg, Germany, 2018.
11. Kubica, P.; Boroń, S.; Czarnecki, L.; Węgrzyński, W. Maximizing the retention time of inert gases used in fixed gaseous extinguishing systems. *Fire Saf. J.* **2016**, *80*, 1–8. [CrossRef]
12. Zhao, H.; Schwabe, A.; Schläfli, F.; Thrash, T.; Aguilar, L.; Dubey, R.K.; Karjalainen, J.; Hölscher, C.; Cristoph, C.; Helbing, D.; et al. Fire evacuation supported by centralized and decentralized visual guidance systems. *Saf. Sci.* **2022**, *145*, 105451. [CrossRef]
13. Seong, G.; Kong, D.J.; Shengzhe, L.; Hakil, K. Fast fire flame detection in surveillance video using logistic regression and temporal smoothing. *Fire Saf. J.* **2016**, *79*, 37–43.
14. Morgan, A. Left Luggage, Automatic Fire Detection and the New Century. *Fire Eng. J.* **2000**, *60*, 37–39.
15. Suproniuk, M.; Paś, J. Analysis of electrical energy consumption in a public utility buildings. *Przegląd Elektrotechniczny* **2019**, *95*, 97–100. [CrossRef]
16. Bernardo, L.; Oliveira, R.; Tiago, R.; Pinto, P. A fire monitoring application for scattered wireless sensor networks: A peer-to-peercross-layering approach. In Proceedings of the International Conference on Wireless Networks and Systems, Barcelona, Spain, 28–31 July 2007; pp. 28–31.
17. Stawowy, M.; Rosiński, A.; Paś, J.; Klimczak, T. Method of Estimating Uncertainty as a Way to Evaluate Continuity Quality of Power Supply in Hospital Devices. *Energies* **2021**, *14*, 486. [CrossRef]
18. Spertino, F.; Amato, A.; Casali, G.; Ciocia, A.; Malgaroli, G. Reliability Analysis and Repair Activity for the Components of 350 kW Inverters in a Large Scale Grid-Connected Photovoltaic System. *Electronics* **2021**, *10*, 564. [CrossRef]

19. Rosiński, A. Reliability analysis of the electronic protection systems with mixed—Three branches reliability structure. In *Reliability, Risk and Safety: Theory and Applications*; Radim, B.C., Guedes, S., Martorell, S., Eds.; CRC Press/Balkema: London, UK, 2010; pp. 1637–1641. [CrossRef]

20. Dhillon, B.S. *Applied Reliability and Quality, Fundamentals, Methods and Procedures*; Springer: London, UK, 2006; p. 186.

21. Chiodo, E.; De Falco, P.; Di Noia, L. Challenges and New Trends in Power Electronic Devices Reliability. *Electronics* **2021**, *10*, 925. [CrossRef]

22. Paś, J.; Klimczak, T. Selected issues of the reliability and operational assessment of a fire alarm system. *Ekspolatacja Niezawodn. Maint. Reliab.* **2019**, *21*, 553–561. [CrossRef]

23. Zajkowski, K. Settlement of reactive power compensation in the light of white certificates. In Proceedings of the E3S Web of Conferences 19, UNSP 01037, Polanica Zdroj, Poland, 13–15 September 2017. [CrossRef]

24. Morgan, A. New fire detection concepts. *Fire Saf. Eng.* **2000**, *7*, 35–37.

25. Roman, D.; Saxena, S.; Bruns, J.; Valentin, R.; Pecht, M.; Flynn, D. A Machine Learning Degradation Model for Electrochemical Capacitors Operated at High Temperature. *IEEE Access* **2021**, *9*, 25544–25553. [CrossRef]

26. Lee, C.; Jo, S.; Kwon, D.; Pecht, M.G. Capacity-Fading Behavior Analysis for Early Detection of Unhealthy Li-Ion Batteries. *IEEE Trans. Ind. Electron.* **2021**, *68*, 2659–2666. [CrossRef]

27. Boroń, S.; Węgrzyński, W.; Kubica, P.; Czarnecki, L. Numerical modelling of the fire extinguishing gas retention in small compartment. *Appl. Sci.* **2019**, *9*, 663. [CrossRef]

28. Krzykowski, M.; Paś, J.; Rosiński, A. Assessment of the level of reliability of power supplies of the objects of critical infrastructure. In Proceedings of the IOP Conference Series: Earth and Environmental Science, Krakow, Poland, 14–17 November 2017; Volume 214, p. 012018. [CrossRef]

29. Da Penha, O.S.; Nakamura, E.F. Fusing light and temperature data for fire detection. In Proceedings of the IEEE Symposium on Computers and Communications (ISCC), Riccione, Italy, 22–25 June 2010; pp. 107–112.

30. Wang, C. *Structural Reliability and Time-Dependent Reliability*; Springer: Singapore, 2021.

31. Pas, J.; Rosinski, A.; Chrzan, M.; Bialek, K. Reliability-Operational Analysis of the LED Lighting Module Including Electromagnetic Interference. *IEEE Trans. Electromagn. Compat.* **2020**, *62*, 2747–2758. [CrossRef]

32. Serio, M.A.; Bonamno, A.S.; Knight, K.S.; Newman, J.S. Fourier Transform Infrared Diagnostics for Improved Fire Detection Systems. In Proceedings of the NIST Annual Conference on Fire Research, Gaithersburg, MD, USA, 28–31 October 1996.

33. So, A.T.P.; Chan, W.L. A computer-vision-based and fuzzy-logic-aided security and fire-detection system. *J. Fire Technol.* **1994**, *30*, 341–356. [CrossRef]

34. Rahardjo, H.A.; Prihanton, M. The most critical issues and challenges of fire safety for building sustainability in Jakarta. *J. Build. Eng.* **2020**, *29*, 101133. [CrossRef]

35. Cadena, J.E.; Osorio, A.F.; Torero, J.L.; Reniers, G.; Lange, D. Uncertainty-based decision-making in fire safety: Analyzing the alternatives. *J. Loss Prev. Process. Ind.* **2020**, *68*, 104288. [CrossRef]

36. Ding, L.; Ji, J.; Khan, F.; Li, X.; Wan, S. Quantitative fire risk assessment of cotton storage and a criticality analysis of risk control strategies. *Fire Mater.* **2020**, *44*, 165–179. [CrossRef]

37. Duer, S.; Zajkowski, K.; Płocha, I.; Duer, R. Training of an artificial neural network in the diagnostic system of a technical object. *Neural Comput. Appl.* **2012**, *22*, 1581–1590. [CrossRef]

38. Wu, H.; Wu, D.; Zhao, J. An intelligent fire detection approach through cameras based on computer vision methods. *Process. Saf. Environ. Prot.* **2019**, *127*, 245–256. [CrossRef]

39. Si, X.-S.; Zhou, D. A Generalized Result for Degradation Model-Based Reliability Estimation. *IEEE Trans. Autom. Sci. Eng.* **2014**, *11*, 632–637. [CrossRef]

40. Stawowy, M.; Perlicki, K.; Sumiła, M. Comparison of uncertainty multilevel models to ensure ITS Services. In *Safety and Reliability—Theory and Applications: Proceedings of ESREL*; Cepin, M., Radim, B., Eds.; CRC Press/Balkema: London, UK, 2017; pp. 2647–2652. [CrossRef]

41. Sharma, A.; Singh, P.K.; Kumar, Y. An integrated fire detection system using IoT and image processing technique for smart cities. *Sustain. Cities Soc.* **2020**, *61*, 102332. [CrossRef]

42. Paś, J. Shock a disposable time in electronic security systems. *J. KONBiN* **2016**, *38*, 5–31. [CrossRef]

43. Shuai, X.; Hao, C.; Feng, D.; Sihang, C. Online sensorless fault diagnosis and remediation strategies selection of transistors for power converter in SRD. *IET Electr. Power Appl* **2019**, *13*, 1553–1564. [CrossRef]

44. Klimczak, T.; Paś, J. Reliability and Operating Analysis of Transmission of Alarm Signals of Distributed Fire Signaling System. *J. KONBiN* **2019**, *49*, 165–174. [CrossRef]

45. Milic, M.; Ljubenovic, M. Arduino-Based Non-Contact System for Thermal-Imaging of Electronic Circuits. In Proceedings of the 2018 Zooming Innovation in Consumer Technologies Conference (ZINC), Novi Sad, Serbia, 30–31 May 2018; pp. 62–67. [CrossRef]

46. Paś, J. *Operation of Electronic Transportation Systems*; Publishing House University of Technology and Humanities: Radom, Poland, 2015.

47. Østrem, L.; Sommer, M. Inherent fire safety engineering in complex road tunnels—Learning between industries in safety management. *Saf. Sci.* **2021**, *134*, 105062. [CrossRef]

48. Stawowy, M. Model for information quality determination of teleinformation systems of transport. In *Safety and Reliability: Methodology and Applications, Proceedings of the 25th European Safety and Reliability Conference, ESREL 2015, Zurich, Switzerland, 10 September 2015*; Nowakowski, T., Młyńczak, M., Jodejko-Pietruczuk, A., Werbińska–Wojciechowska, S., Eds.; CRC Press/Balkema: London, UK, 2015; pp. 1909–1914.
49. Rosiński, A. *Modelling the Maintenance Process of Transport Telematics Systems*; Publishing House Warsaw University of Technology: Warsaw, Poland, 2015.
50. Paś, J.; Klimczak, T. Modeling of the process of selected fire signaling systems. *Diagnostyka* **2019**, *20*, 81–88. [CrossRef]
51. Joglar, F. Reliability, Availability, and Maintainability. In *SFPE Handbook of Fire Protection Engineering*; Hurley, M., Ed.; Springer: New York, NY, USA, 2016; pp. 2875–2940.
52. Tostado-Véliz, M.; Bayat, M.; Ghadimi, A.A.; Jurado, F. Home energy management in off-grid dwellings: Exploiting flexibility of thermostatically controlled appliances. *J. Clean. Prod.* **2021**, *310*, 127507. [CrossRef]
53. Hulida, E.; Pasnak, I.; Koval, O.; Tryhuba, A. Determination of the Critical Time of Fire in the Building and Ensure Successful Evacuation of People. *Period. Polytech. Civ. Eng.* **2019**, *63*, 308–316. [CrossRef]
54. Kaniewski, P.; Smagowski, P.; Konatowski, S. Ballistic Target Tracking with Use of Cinetheodolites. *Int. J. Aerosp. Eng.* **2019**, *2019*, 3240898. [CrossRef]
55. Idris, A.M.; Rusli, R.; Burok, N.A.; Nabil, N.H.M.; Ab Hadi, N.S.; Karim, A.H.M.A.; Ramli, A.F.; Mydin, I. Human factors influencing the reliability of fire and gas detection system. *Process. Saf. Prog.* **2020**, *39*, e12116. [CrossRef]
56. Menon, S.; Chen, D.Y.; Osterman, M.; Pecht, M.G. Copper Trace Fatigue Life Modeling for Rigid Electronic Assemblies. *IEEE Trans. Device Mater. Reliab.* **2021**, *21*, 79–86. [CrossRef]
57. Paś, J.; Rosiński, A.; Wiśnios, M.; Majda-Zdancewicz, E.; Łukasiak, J. *Electronic Security Systems. Introduction to the Laboratory*; Military University of Technology: Warsaw, Poland, 2018.
58. Mahdipour, E.; Dadkhah, C. Automatic fire detection based on soft computing techniques: Review from 2000 to 2010. *Artif. Intell. Rev.* **2012**, *42*, 895–934. [CrossRef]
59. Kozłowski, E.; Borucka, A.; Świderski, A. Application of the logistic regression for determining transition probability matrix of operating states in the transport systems. *Ekspolatacja Niezawodn. Maint. Reliab.* **2020**, *22*, 192–200. [CrossRef]
60. Duer, S. Assessment of the Operation Process of Wind Power Plant's Equipment with the Use of an Artificial Neural Network. *Energies* **2020**, *13*, 2437. [CrossRef]
61. Krepl, V.; Shaheen, H.I.; Fandi, G.; Smutka, L.; Muller, Z.; Tlustý, J.; Husein, T.; Ghanem, S. The Role of Renewable Energies in the Sustainable Development of Post-Crisis Electrical Power Sectors Reconstruction. *Energies* **2020**, *13*, 6326. [CrossRef]
62. Major, S.; Frickenstein, G. *Reliability Theory with Applications to Preventive Maintenance*; Llya Gertsbakh Springer: Berlin/Heidelberg, Germany, 2002; pp. 1111–1113; ISBN 3-540-67275-3. [CrossRef]
63. Duer, S.; Bernatowicz, D.; Wrzesień, P.; Duer, R. The diagnostic system with an artificial neural network for identifying states in multi-valued logic of a device wind power. In *Communications in Computer and Information Science*; Springer: Berlin, Germany, 2018; Volume 928, pp. 442–454.
64. Arias, S.; La Mendola, S.; Wahlqvist, J.; Rios, O.; Nilsson, D.; Ronchi, E. Virtual Reality Evacuation Experiments on Way-Finding Systems for the Future Circular Collider. *Fire Technol.* **2019**, *55*, 2319–2340. [CrossRef]
65. Rosiński, A.; Paś, J.; Łukasiak, J.; Szulim, M. Exploitation of electronic systems in building objects exposed to impact of strong electromagnetic pulses. In *Proceedings of the 29th European Safety and Reliability Conference (ESREL), Hannover, Germany, 22–26 September 2019*; Beer, M., Zio, E., Eds.; Research Publishing Services: Singapore, 2019; pp. 3320–3332.

Article

The Issue of Operating Security Systems in Terms of the Impact of Electromagnetic Interference Generated Unintentionally

Krzysztof Jakubowski [1], Jacek Paś [2] and Adam Rosiński [3,*]

[1] National Cyber Security Centre, Gen. Buka 1, 05-119 Legionowo, Poland; krzysztof.jakubowski.wel@gmail.com
[2] Division of Electronic Systems Exploitations, Faculty of Electronics, Institute of Electronic Systems, Military University of Technology, 2 Gen. S. Kaliski St, 00-908 Warsaw, Poland; jacek.pas@wat.edu.pl
[3] Division Telecommunications in Transport, Faculty of Transport, Warsaw University of Technology, 75 Koszykowa St, 00-662 Warsaw, Poland
* Correspondence: adam.rosinski@pw.edu.pl

Abstract: This article discusses issues regarding electromagnetic interference generated unintentionally by transport telematics systems and electronic security systems (ESS) located within a railway area. These systems should operate correctly, since they ensure the safety of both vehicles and passengers. The electronic devices they use are exposed to electromagnetic interference that may lead to incorrect ESS functioning. In order to determine the impact of electromagnetic interference on ESS, the authors measured unintentional low-frequency electromagnetic field generated by MV—15 and 30 kV—power lines. This enabled determining the areas with maximum values of electromagnetic interference. The next stage of the research was to develop an ESS operating process model that takes into account the impact of unintentionally generated electromagnetic interference on the operating process. Introducing the electromagnetic interference impact coefficient enables a rational selection of solutions aimed at protecting against electromagnetic interference through the application of technical and organizational measures.

Keywords: electronic security system; electromagnetic interference; operation

1. Introduction

Transport telematics systems and electronic security systems (ESS) located in a metal container on a railway route are considered to be the "State's critical infrastructure items" [1–3]. Their most important task is to ensure the security of vehicle movement on railway routes [4,5]. They are located in a special, structurally reinforced metal container and are protected against the action of external and internal destructive factors. ESS elements can also be found inside buildings and in a publicly available external environment, where they are exposed to variable environmental parameters—temperature [6], humidity, etc., as well as the unintentional low-frequency electromagnetic field generated by medium voltage—15, 30 kV—power lines. These lines are a supply source for the aforementioned systems (Figure 1) [7].

Citation: Jakubowski, K.; Paś, J.; Rosiński, A. The Issue of Operating Security Systems in Terms of the Impact of Electromagnetic Interference Generated Unintentionally. *Energies* **2021**, *14*, 8591. https://doi.org/10.3390/en14248591

Academic Editors: Anna Richelli and Stanisław Duer

Received: 5 November 2021
Accepted: 14 December 2021
Published: 20 December 2021

Publisher's Note: MDPI stays neutral with regard to jurisdictional claims in published maps and institutional affiliations.

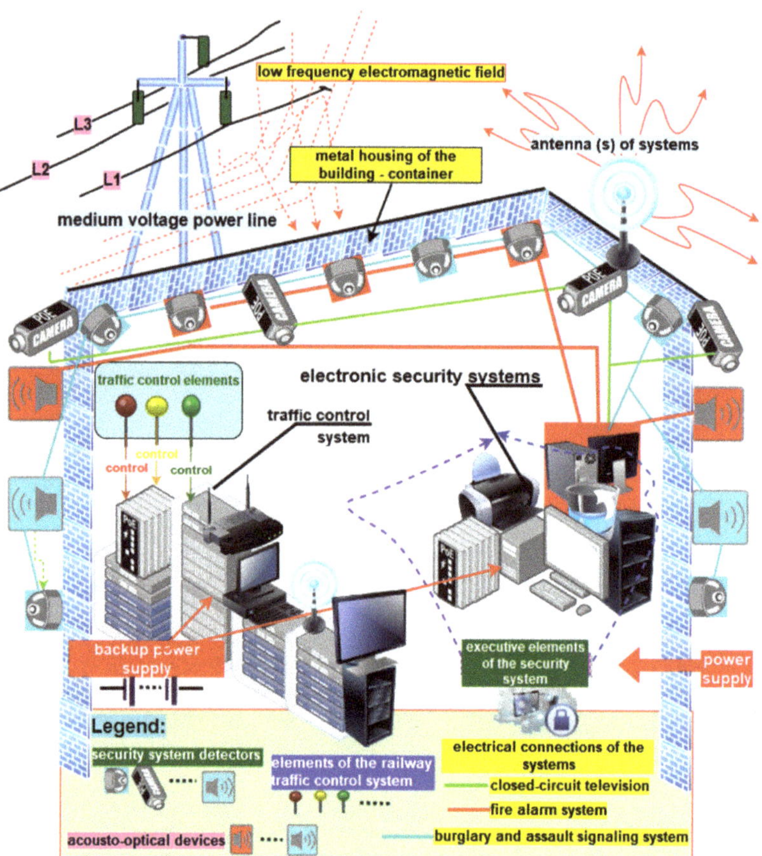

Figure 1. Arrangement of electronic security systems in a metal container due to their important functions within the State's communication structure. Designations: L1, L2, L3—individual voltage phases found in the overhead power line supplying electricity to transport telematics systems and electronic security systems. The power is supplied from the MV grid and from backup power sources (battery banks for electronic security systems and UPSs for rail transport telematics systems due to the rated power demand).

Transport telematics systems and electronic security systems are supplied from the 15 kV or 30 kV medium-voltage grid. The overhead power line is 5 or 10 m away from the container where the aforementioned electronic systems are located, in accordance with applicable national standards that apply to ensuring protection against the impact of unintentional electromagnetic field on living and inanimate organisms. A transformer located outside the container ensures an appropriate voltage level—230 V electronic power supplies. Backup power supply in the form of battery banks for electronic security systems is provided in order to guarantee the high reliability of telematics systems and ESS in the container. It consists of a fire alarm system (FAS) [8,9], an intrusion detection system (IDS), and closed-circuit television (CCTV). Battery banks should exhibit sufficient capacity that is determined in accordance with applicable standards (taking into account detection and alarm currents that should be ensured for a specified operation period of these systems). Due to the higher consumption of working currents by telematics systems, the backup power supply is provided within such a technical facility by UPS devices. Both battery banks and UPS devices are technically organized in a manner sufficient for the purposes of guaranteeing an adequate operational safety level by providing power supply continuity. In order to ensure an adequate power supply reliability level, designers use, e.g., redundancy—cold, warm, and hot reserve or the so-called "failsafe" principle. Electronic security system elements are operated inside the structurally and mechanically reinforced metal container, where an air-conditioning system located outside the structure provides a strictly defined temperature and humidity. In order to secure appropriate protection against external actions that may impact the technical facility in question, selected ESS elements are placed on the container walls. These include, among others, cameras, IDS detectors, fire detectors, and signaling and acoustic devices. These signaling and audio devices are distinctive, separately announcing a fire and intrusion alarm. IDS and CCTV systems are integrated in order to maintain an appropriate security level. The FAS is not subject to integration with the a/m systems due to the legislation applicable in Poland. Information on all breaches, alarms, damage, activations, deactivations, etc. related to all safety or fire zones is saved in the nonvolatile memory of the alarm control units of the aforementioned devices. Furthermore, the information is sent to a remote receiving center (ARC) via two independent telecommunication routes—hardwired and radio [10–12]. In addition, the FAS has a separate, standalone alarm and damage signal transmission device (ADSTD), where the alarm signal is forwarded to the State Fire Service (PSP). All security systems, because of the function performed within a critical infrastructure facility, have backup alarms and damage signal transmitter. There are no people in the container who are responsible for service and repairs, and who are on continuous duty—24 h per day. Information on alarm and damage is sent to an ARC for security systems, whereas, in the case of transport telematics systems, it is sent to a separate technical point, which monitors the operating process involving a given railway route section. The container's layout and position in the vicinity of medium-voltage power lines necessitates including the impact of unintentional low-frequency electromagnetic radiation in the ESS operating process. Due to the methodology of the low-frequency measurements, two separate electromagnetic field components—E electric field strength and B magnetic field induction—need to be taken into account separately. An electromagnetic field unintentionally generated by an MV power line may adversely impact the operating process of the aforementioned electronic systems. A variable unintentional electric or magnetic field may lead to the generation of interference inside such electronic systems (individual elements—detectors, control or power modules, alarm control units, etc.), as well as power, signal, or control cables, transmission buses, and detection circuits in the case of FAS [13–15].

A previous study [16] presented issues associated with the use of PV module energy in rail transport and its storage in order to supply vehicles. Such solutions are ecofriendly, but require the use of electrical and electronic devices, which also impact electronic security systems. For this reason, the analysis of the applied solutions aimed at improving resistance to electromagnetic interference is important.

Moreover, the issues regarding radio transmission are significant for the correct functioning of electronic security systems. The authors of [17] described the issue associated with high-strength electromagnetic fields generated by high-power pulse transmitters. This may severely affect electrical and electronic systems, unless rational solutions improving the strength and resistance of ESS to such interference are applied.

The location of the electromagnetic interference source is also important. Issues in this field were discussed in [18]. The authors suggested an innovative localization technique based on the time-reversal cavity (TRC) concept. This is an interesting approach; nevertheless, due to the measurement specificity and the available devices for measuring electromagnetic fields, the authors applied a traditional approach for two MV power lines in this article.

The authors of [19] described the problem of energy recovery upon train braking with regard to electromagnetic compatibility (more precisely—the interference between generated current harmonics and the rail signaling system). The conducted simulations enabled suggesting an asymmetric brake control in order to reduce current harmonics.

The issue regarding the impact of electromagnetic interference on the functioning of electronic rail transport devices is of particular importance in the case of high-speed railway. A previous study [20] presented an original approach to assessing the impact of electromagnetic interference on radio transmission that is based on common EMI characteristics. The application of radio transmission using local coding enables maintaining correct communication parameters.

The authors of [21] also addressed the interactions actions of the electromagnetic environment on high-speed railway. A graph model was developed in order to analyze the impact of electromagnetic interference on the operation of electronic railway equipment. It enables a more detailed reflection of the relationships for individual subsystems in the context of electromagnetic compatibility.

Despite the presented various approaches, the electromagnetic compatibility analysis involving electronic devices lacks studies that would link reliability and operational modeling of electronic railway equipment with resistance to electromagnetic interference. For this purpose, the authors of this article studied the issues in this field.

The issue of powering transport telematics systems and electronic security systems is characterized at the beginning of the paper. Particular attention is paid to the reliability and operational aspects, and the impact of unintentional low-frequency electromagnetic radiation. Next, the authors present the results of tests involving the electromagnetic environment surrounding medium-voltage lines supplying a container security system. This was used as a base to conclude that it was justified to apply solutions resulting in mitigated low-frequency electromagnetic field impact on the systems in question. Another stage of the deliberations by the authors was the analysis of the security system operating process in terms of the impact of unintentionally generated electromagnetic interference. The determined relationships enable rationally selecting solutions aimed at protecting systems against electromagnetic interference.

2. Study of the Electromagnetic Environment Surrounding Medium-Voltage Lines Supplying a Container Security System

The study was conducted for two medium-voltage (15 kV and 30 kV) power lines that supplied a metal container housing transport telematics systems and electronic security systems. The aforementioned containers were located near a railway communication route; however, the significant distance (over 250 m) from the overhead contact line allowed the authors to discard unintentional, nonstationary electromagnetic interference generated by the moving electric locomotives [22,23]. Operating electrical or electronic devices that are powered by electricity leads to generating electromagnetic, electric, or magnetic fields. A field generated artificially, intentionally or unintentionally by humans, deforms the natural electromagnetic environment. Most often, humans encounter in their nearest surroundings (their workplace or residence) unintentional electromagnetic radiation sources that are primarily of the low-frequency range. Power transmission lines are one of the main sources of unintentional electromagnetic radiation. Due to their coverage (overhead lines in particular), they emit electromagnetic radiation over a large area. Two frequency subranges for field meters were introduced in the measurement methodology involving the issues associated with diagnosing low-frequency electromagnetic fields, e.g., sources with a frequency of 50 Hz that are generated by power lines. The first range included extremely low frequencies (ELF)—from 5 Hz to 2 kHz, while the second involved very low frequencies (VLG)—from 2 kHz to 100 kHz. The space around the medium-voltage lines was diagnosed within these frequencies, pursuant to the applicable regulations on the measurement methodology. In order to investigate the impact of medium-voltage lines on the electromagnetic environment inside and outside the container with installed transport telematics systems, the authors measured the distribution of electromagnetic field components under three power lines supplying technical facilities, i.e., a 30 kV line and two different 15 kV lines. The measurements were taken at a positive outside temperature 20–21 °C and humidity 38–42%. The so-called "electromagnetic field background measurements" were conducted prior to commencing the measurements covering the electromagnetic field generated by the studied power lines. They were taken in places where the natural electromagnetic field of the Earth is undistorted. The results are listed in Table 1.

Table 1. Mean magnetic field induction B and electric field strength E values in the ELF and VLF frequency ranges near the studied lines—field background measurement.

	30 kV Line	15 kV Line No. 1	15 kV Line No. 2
Mean magnetic field induction B, ELF range	0.02 µT	0.02 µT	0.015 µT
Mean magnetic field induction B, VLF range	0.6 nT	0.4 nT	0.4 nT
Mean electric field strength E, ELF range	46 V/m	26 V/m	23 V/m
Mean electric field strength E, VLF range	2.2 V/m	2.3 V/m	2.2 V/m

Studies involving the impact of unintentional electromagnetic field on the external and internal environment of a container housing telematics and security systems were conducted in accordance with Figure 2.

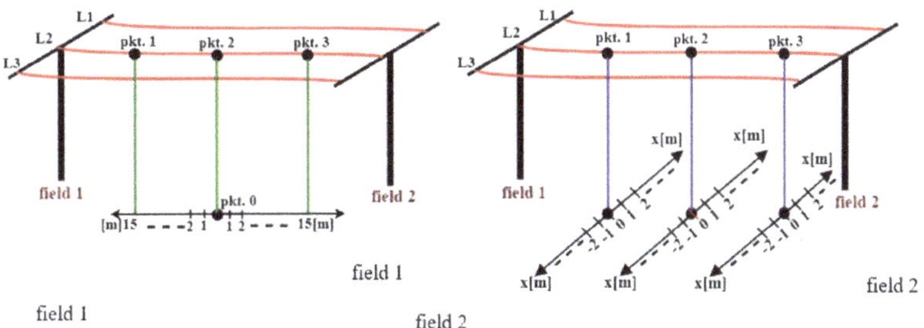

Figure 2. Diagram of a test stand for measuring electromagnetic field components along the line's phase conductors, and across the medium-voltage power line.

The presented characteristic curve (Figure 3) shows a significant decline in magnetic field induction B with increased distance from the studied lines. Magnetic field induction B reaches a field background level at a distance of 17–19 m from point 0 in the case of the 15 kV power line No. 1, whereas, for the 15 kV power line No. 2 and the 30 kV line, this distance is 7–9 m from point 0, with the measurement conducted in accordance with Figure 2. Disproportions between the maximum and minimum magnetic field induction B, depending on the measurement location, can be noticed when analyzing the resulting distribution. According to the Biot–Savart law, magnetic field induction B is proportional to the value of the flowing current and inversely proportional to the distance of the measuring point from the field source. Higher voltage of a transmission line supplying a container does not necessarily mean the presence of higher magnetic field induction B values. This is confirmed by the characteristic curves in Figure 3. The values of magnetic field induction B components B_x, B_y, and B_z are shown in Figure 4 for measuring point No. 2 (Figure 2), due to the fact that it provided the highest magnetic field induction B values.

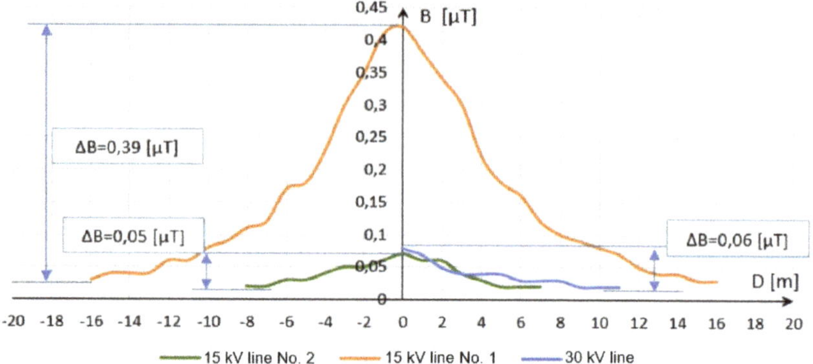

Figure 3. Distribution of magnetic field induction B across the line, for measuring point No. 1 and the ELF frequency range (induction measuring methodology as in Figure 2).

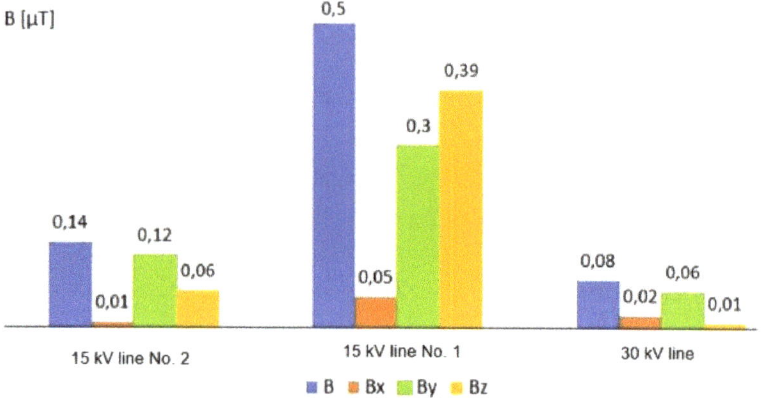

Figure 4. The value of magnetic field induction B and its components B_x, B_y, B_z at the point of maximum cable sag in the power line supplying the container with telematics and security systems. Maximum magnetic field induction B under MV lines for the ELF range was recorded at measuring point No. 2.

Magnetic induction in the VLF range was also measured at measuring point No. 2 (Figure 2). The results are shown in Figure 5.

The maximum magnetic field induction B for the VLF range under the 30 kV line was 2.2 nT, which was achieved at a distance of 5 m from point 0 (measurement in accordance to Figure 2). In contrast, the maximum magnetic field induction B was 0.9 nT for 15 kV line No. 1 and 0.7 nT for 15 kV line No. 2. Figures 6 and 7 show the measurement of magnetic induction along phase conductors of the container supply lines for the ELF range and two supply voltages—15 and 30 kV. The measurement procedure is shown in Figure 2.

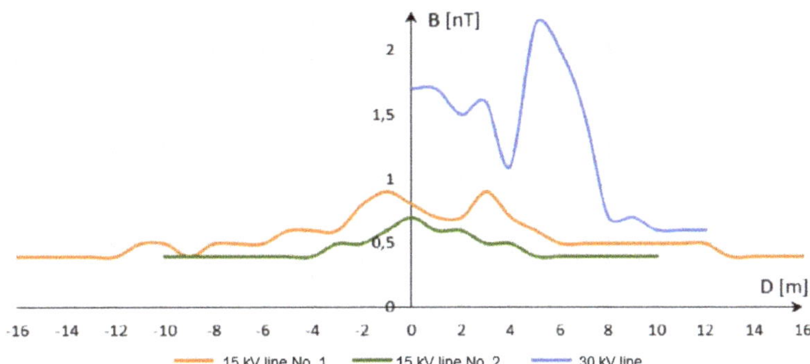

Figure 5. Distribution of magnetic field induction B across the line for measuring point No. 2, and the VLF frequency range (induction measuring methodology as in Figure 2).

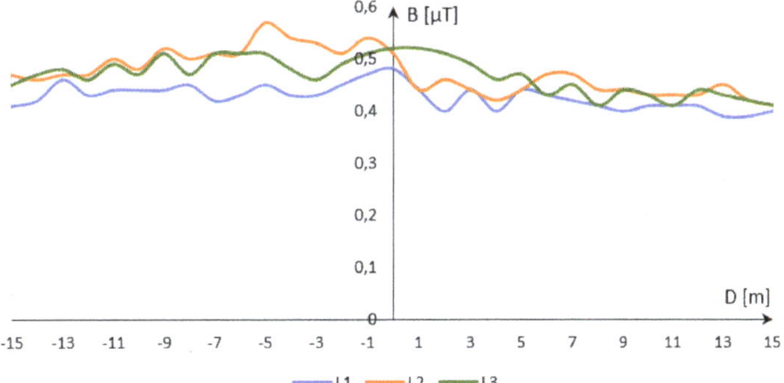

Figure 6. Distribution of magnetic field induction B along L1, L2, L3 phase conductors of the 15 kV power line No. 1, ELF frequency range.

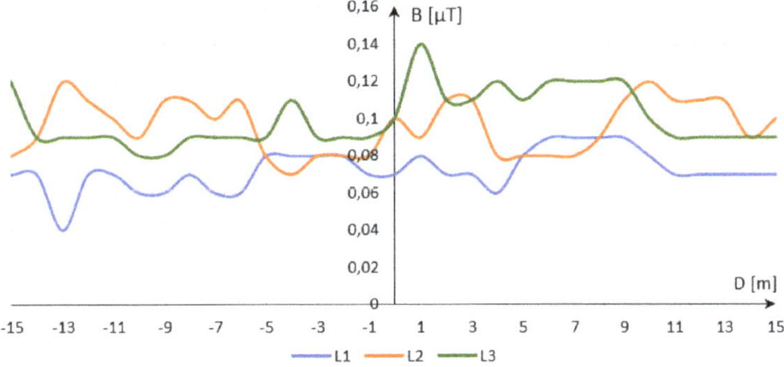

Figure 7. Distribution of magnetic field induction B along L1, L2, L3 phase conductors of the 30 kV power line No. 1, ELF frequency range.

The maximum magnetic field induction B measured along the phase conductors for the ELF range was identified under the central L2 conductor of the 15 kV line No. 1 and amounted to 0.57 µT, while the value under the 30 kV line was equal to 0.14 µT. Figures 8 and 9 show the magnetic field induction B along phase conductors for the VLF phase (measurement methodology shown in Figure 2). Upon observing the changes in the magnetic field induction B along the phase conductors over the VLF range, it can be noted that the 30 kV supplying the container was distinguished by higher values relative to both

15 kV lines. There were also significant differences between the maximum and minimum magnetic field induction B value depending on the rated voltage of the line supplying the container with telematics and security systems.

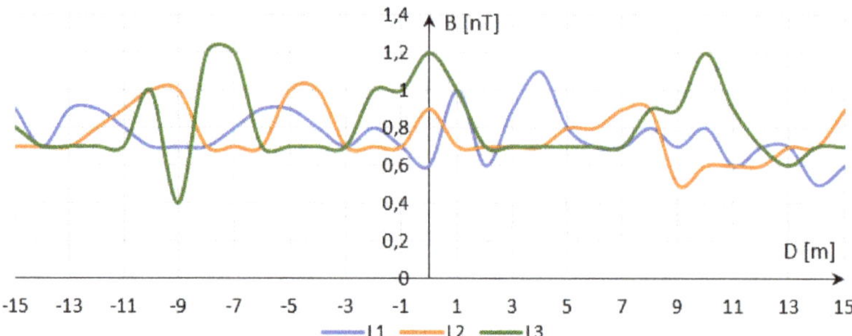

Figure 8. Distribution of magnetic field induction B along L1, L2, L3 phase conductors of the 15 kV container power line No. 1, VLF frequency range.

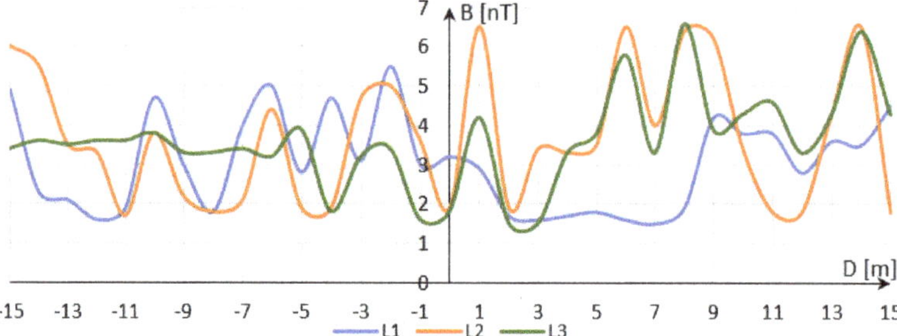

Figure 9. Distribution of magnetic field induction B along L1, L2, L3 phase conductors of the 30 kV container power line, VLF frequency range.

Changes in the magnetic field induction B over time were also measured. The measurements were taken in the location where the highest values were recorded. The magnetic field induction B value was read every minute for 15 min. Measurement results are shown in Figure 10. On comparing the magnetic field induction B values obtained over time, it can be concluded that they remained on average at a constant level. It was 0.53 µT for the 15 kV line No. 1 and 0.08 µT for the 30 kV and 15 kV No. 2 lines.

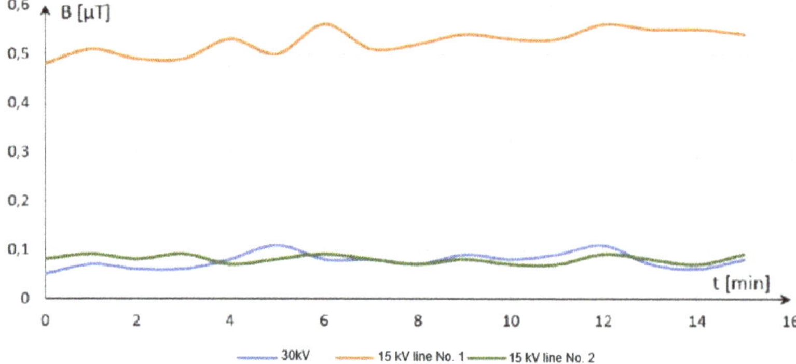

Figure 10. Magnetic field induction B change fluctuations under medium-voltage lines for the ELF range.

Figures 11 and 12 show the distribution of electric field strength E across MV power lines supplying the container with transport telematics system and ESS, at two measuring points, in accordance with Figure 2.

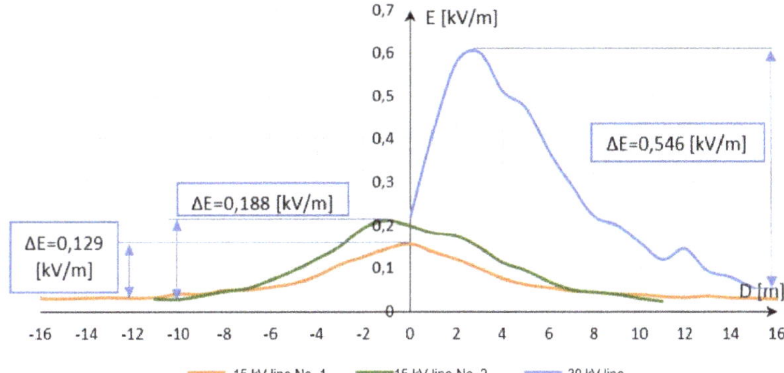

Figure 11. Distribution of electric field strength E across the line, for measuring point No. 1 and the ELF frequency range (induction measuring methodology as in Figure 2).

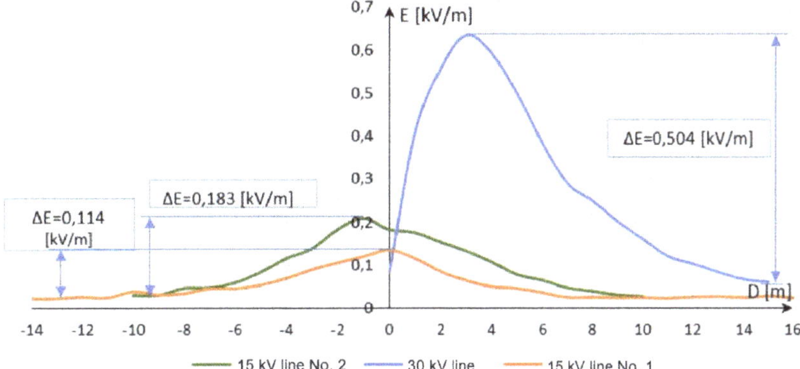

Figure 12. Distribution of electric field strength E across the line, for measuring point No. 3 and the ELF frequency range (induction measuring methodology as in Figure 2).

Electric field strength E decreased with increasing distance from point 0 and reached a background value at a distance of 11–13 m for the 15 kV power line and 15–16 m for the 30 kV line. As in the case of the magnetic field induction B, the electric field strength E also reached the highest values at measuring point No. 2. The 30 kV power line, however, achieved significantly higher electric field strength E values. Along with the increase of the line voltage, the electric field strength E also increased.

The electric field strength E was also measured in terms of the VLF range for measuring point No. 2 (measurement methodology as in Figure 2). Measurement results are shown in Figure 13.

In the course of analyzing the characteristic curves, it can be concluded that 15 kV lines produced virtually no electric field strength E component over the VLF range. The changes in the values over the entire measuring range fluctuated from 2.1 to 2.3 V/m, which corresponded to the values of the measured electric field background. The changes in the electric field strength E along the phase conductors of container-supplying MV power lines were also analyzed. The results of these tests are shown in Figures 14 and 15.

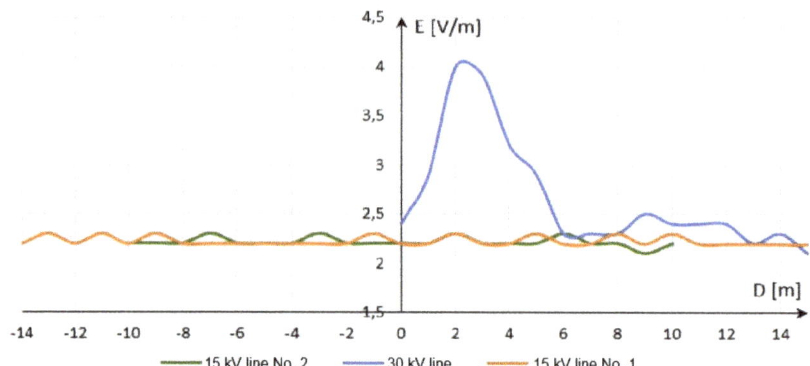

Figure 13. Distribution of electric field strength E across the line, for measuring point No. 2 and the VLF frequency range (induction measuring methodology as in Figure 2).

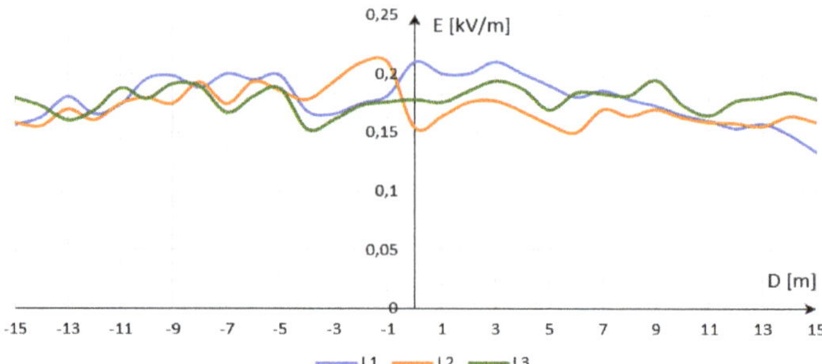

Figure 14. Distribution of electric field strength E along L1, L2, L3 phase conductors of the 15 kV container power line, ELF frequency range (measurement methodology as in Figure 2).

Figure 16 shows changes in the electric field strength E under MV lines supplying the container, for the ELF range. In the case of both 15 kV lines, the electric field strength could be considered constant, and its mean value was 0.19 kV/m and 0.21 kV/m for 15 kV lines No. 1 and No. 2, respectively. In the case of the 30 kV line, the electric field strength ranged from 0.53 kV/m to 0.73 kV/m.

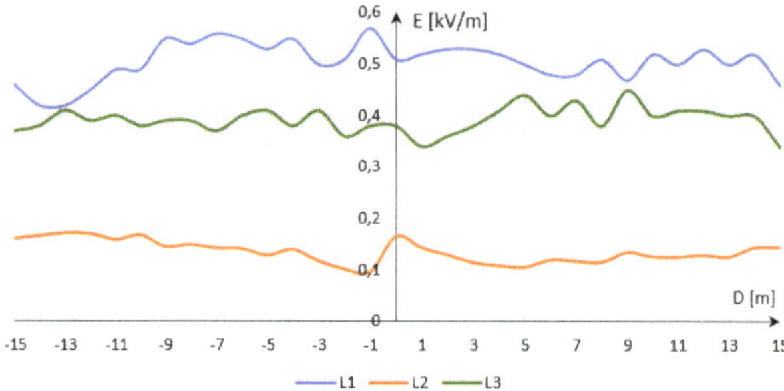

Figure 15. Distribution of electric field strength E along L1, L2, L3 phase conductors of the 30 kV container power line, ELF frequency range (measurement methodology as in Figure 2).

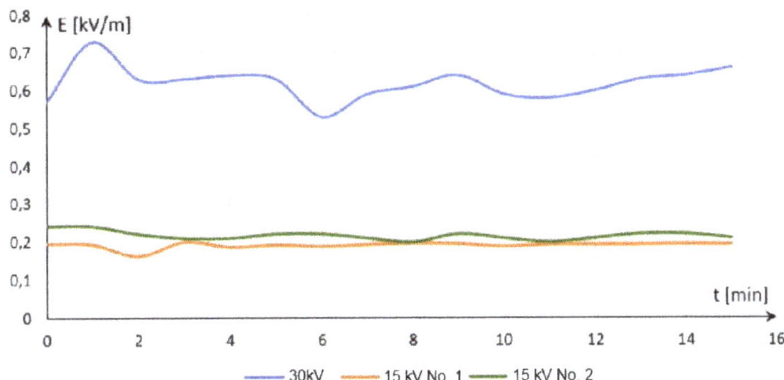

Figure 16. Electric field strength E fluctuations under MV lines supplying the container, for the ELF range (measurement methodology as in Figure 2).

When analyzing the obtained measurement results for all power lines supplying the container, it can be concluded that the highest magnetic field induction B was recorded under the 15 kV line No. 1 and amounted to 0.57 µT. In contrast, the maximum electric field strength E component value was at a level 0.64 kV/m, which was achieved near the 30 kV line.

3. Metal Container Housing: Determining the Shielding Effectiveness for a Low-Frequency Electromagnetic Field

When constructing and operating electronic, transport telematics, or security equipment and systems, it is required to protect a given technical facility against the impact of undesirable electromagnetic interference over a wide spectrum range [24,25]. Appropriately manufactured guards, usually metallic, called shields are used in such cases in accordance with the electromagnetic compatibility pyramid [26–28]. An electromagnetic shield is always a conductive shield that separates two areas. One is the source, e.g., of an unintentional electromagnetic field, i.e., MV power lines supplying the container in this case. The other does not contain the source, i.e., it is the container interior. This separation applies not only to fields, but also to fault currents. A shield also provides a potential reference for cables routed from the container housing and for electronic circuits located inside the metal container. The most important parameter that characterizes shielding execution is the so-called "shield effectiveness S coefficient". It is a ratio of the shield attenuation determined on the basis of field residual strength (electric, magnetic, or electromagnetic field) measured in the presence of a shield and the field strength measured without the shield. The shielding effectiveness S coefficient is a dimensionless value. Shield attenuation S is most usually expressed in dB, as per Equation (1).

$$S = 20 \cdot \log \cdot (P_{be}/P_{ze}), \tag{1}$$

where P_{be} is the unshielded field strength, and P_{ze} is the shielded field strength.

Shield attenuation is defined in relation to an electrical (Equation (2)), magnetic (Equation (3)), or coupled field. Shield effectiveness relative to the electrical or magnetic field is expressed by Equations (2) and (3), respectively.

$$S_E = 20 \cdot \log \cdot (E_{be}/E_{ze}), \tag{2}$$

$$S_H = 20 \cdot \log \cdot (H_{be}/H_{ze}), \tag{3}$$

where E_{be}, H_{be} are, respectively, unshielded electric and magnetic strengths, and E_{ze}, H_{ze} are, respectively, shielded electric and magnetic strengths.

A positive attenuation values is expressed in *dB* and means a reduction in field strength. A negative attenuation implies a strength increase. In any given electronic device, at various frequency ranges, poorly executed shielding may behave like a directional antenna. Shielding effectiveness is a function with numerous variables and depends on parameters such as frequency, shield geometric structure, location within the shield where the field is determined, type of shielded field—electric, magnetic, or electromagnetic, incidence direction, and wave polarization. The total shielding effectiveness of the material used to manufacture the container where the telematics and security systems are operated is equal to the total relative measures of attenuation losses *A*, deflection losses *R*, and the correction factor *B*. The correction factor *B* takes into account the so-called "numerous deflections in thin shields" (Figure 17). In such a case, the total shielding effectiveness of a metal container can be expressed using Equation (4).

$$S[dB] = A + R + B. \tag{4}$$

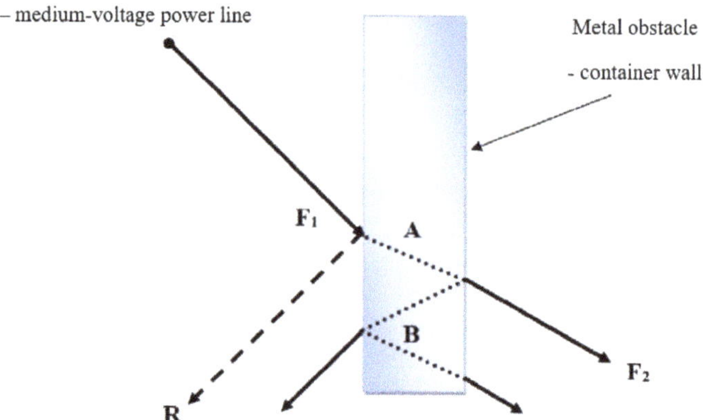

Figure 17. Principle of shielding an electromagnetic wave generated by an MV power line through a metal obstacle constituting the container.

The following materials are sufficient for the shielding of the low-frequency electromagnetic fields that are generated by MV power lines:

- thin, highly conductive materials, 20 to 50 μm thick, since they guarantee high losses related to deflection of electric waves;
- several millimeter thick materials, when shielding a magnetic field. This is brought about by the poor attenuation that is associated with waves deflecting from the shield surface; hence, it must provide high absorption losses if the shield is to be effective (shielding fields with a frequency up to 500 kHz requires the application of a magnetically soft material, e.g., iron with relatively conductivity $\mu_r = 200$ or Alu-metal with a thickness >0.5 mm).

The simplest way to reduce the impact of an electromagnetic radiation source on a given electronic device or element is placing it as far from the source as possible—in this case, away from the medium-voltage supply lines. This will enable locating it in areas where the unintentionally generated electromagnetic field exhibits low values. However, it is impossible to always apply this principle. In such a case, a solution is to mitigate the impact of an electromagnetic field by shielding the source or a selected space around the MV power lines. Shield effectiveness was measured for a metal housing of a container made of sheet metal with thickness h = 1.5 mm. The results of shielding effectiveness measurements for the ELF and VLF ranges are shown in Table 2. Figure 18 shows the shielding effectiveness of a 1.5 mm thick steel material for magnetic field induction B and electric field strength E over the ELF and VLF ranges. The shielding effectiveness value should be taken into account when determining operating process safety indices for security and telematics systems operated inside any metal container.

Table 2. Magnetic field induction B and electric field strength E shielding effectiveness over the described ELF and VLF ranges for individual shields (shield is the metal container housing, with a sheet thickness h—1.5 mm).

Shielding Effectiveness S	S (dB)			
	Magnetic Field Induction B		Electric Field Strength E	
Frequency range	ELF	VLF	ELF	VLF
Shield made of steel sheet, thickness h = 1.5 mm	8.26	13.5	3.57	8.1

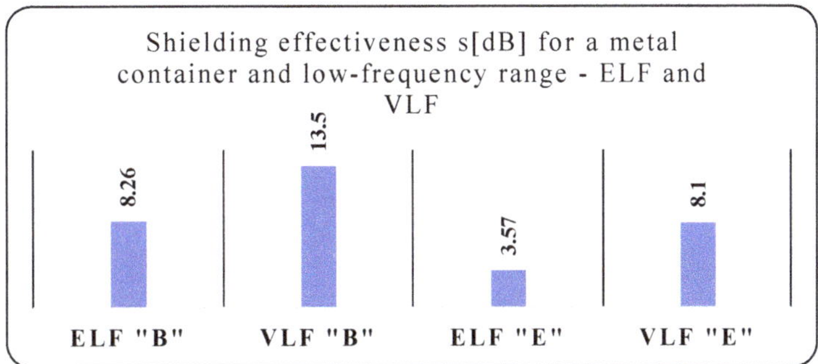

Figure 18. Shielding effectiveness for a metal material with a thickness h = 1.5 mm, for magnetic field induction B and electric field strength E, over the ELF and VFL ranges.

4. Analysis of the Security System Operating Process in Terms of the Impact of Unintentionally Generated Electromagnetic Interference

The impact of use [29], operation and power supply [30,31] conditions, including the impact of unfavorable external [32–35], and internal [36] environmental factors (e.g., low-frequency electromagnetic field) on electronic telematics [37–40] and security systems can be described through, e.g., a change in the intensity of damage to λ components making up a specific system [41,42]. Telematics and security systems and equipment operated in the container contain a large number of active and passive elements. These elements exhibit various margins of resistance to different interference, e.g., electromagnetic, environmental, electrical, thermal, and mechanical, generated in a container in an intentional or unintentional manner. The resistance value undergoes constant change impacted by various environmental factors, e.g., electromagnetic interference, temperature, humidity, and precipitation. Therefore, it is essential to develop an operating process model that takes into account the resistance to electromagnetic interference.

A functional analysis of an electronic security system located in a metal container, taking into account the impact of unintentionally generated electromagnetic interference, enables illustrating the relations therein from the reliability and operational perspective, as demonstrated in Figure 19.

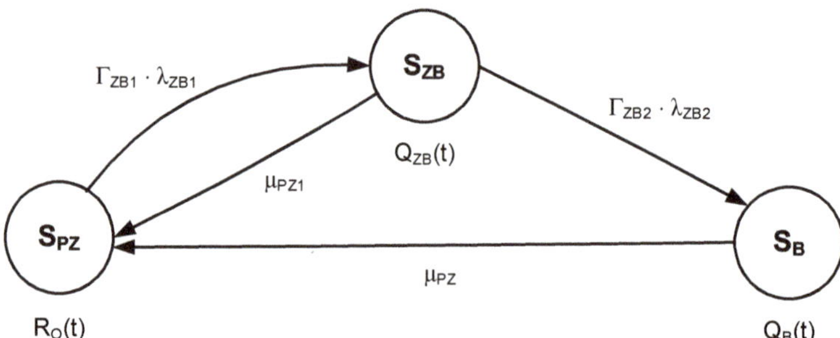

Figure 19. Functional relations within an electronic security system, taking into account the impact of unintentionally generated electromagnetic interference. Designations in Figure: $R_O(t)$—probability function for an ESS to stay in state S_{PZ}; $Q_{ZB}(t)$—probability function for an ESS to stay in state S_{ZB}; $Q_B(t)$—probability function for an ESS to stay in state S_B; λ_{ZB1}—intensity of transitions from state S_{PZ} to state S_{ZB}; μ_{PZ1}—intensity of transitions from state S_{ZB} to state S_{PZ}; λ_{ZB2}—intensity of transitions from state Q_{ZB} to state Q_B; μ_{PZ}—intensity of transitions from state Q_B to state S_{PZ}; Γ_{ZB1} and Γ_{ZB2}—electromagnetic interference impact coefficients.

The state of full fitness S_{PZ} is a state in which an electronic security system operates correctly. The state of security hazard S_{ZB} is a state in which an ESS operates partially correctly. The state of security unreliability (failure) S_B is a state in which an ESS does not operate correctly.

In the event of the ESS staying in a state of full fitness S_{PZ} and the presence of unintentionally generated electromagnetic interference resulting in the inability to perform certain function, the system will shift to a state of security hazard S_{ZB}, with an intensity equal to the product of $\Gamma_{ZB1} \lambda_{ZB1}$.

An ESS in a state of security hazard S_{ZB} can move to a state of full fitness S_{PZ}, if the ability to implement all ESS functions is restored, or security unreliability Q_B, with an intensity equal to the product of $\Gamma_{ZB2} \lambda_{ZB2}$ in the event of electromagnetic interference, resulting in ESS's inability to perform its functions.

Transition from a state of security unreliability S_B to a state of full fitness S_{PZ} takes place when the ability to perform all ESS functions is restored.

The Γ_{ZB1} and Γ_{ZB2} coefficients quantify the impact of unintentionally generated electromagnetic interference, taking into account the applied countermeasures.

The electronic security system presented in Figure 19 is described by the following Kolmogorov–Chapman system of equations:

$$\begin{aligned} R_0'(t) &= -\Gamma_{ZB1} \cdot \lambda_{ZB1} \cdot R_0(t) + \mu_{PZ1} \cdot Q_{ZB}(t) + \mu_{PZ} \cdot Q_B(t), \\ Q_{ZB}'(t) &= \Gamma_{ZB1} \cdot \lambda_{ZB1} \cdot R_0(t) - \mu_{PZ1} \cdot Q_{ZB}(t) - \Gamma_{ZB2} \cdot \lambda_{ZB2} \cdot Q_{ZB}(t), \\ Q_B'(t) &= \Gamma_{ZB2} \cdot \lambda_{ZB2} \cdot Q_{ZB}(t) - \mu_{PZ} \cdot Q_B(t). \end{aligned} \tag{5}$$

If we assume the following baseline conditions for further analysis:

$$\begin{aligned} R_0(0) &= 1, \\ Q_{ZB}(0) &= Q_B(0) = 0, \end{aligned} \tag{6}$$

and then apply the Laplace transform [43–45], we obtain the following set of linear equations:

$$\begin{aligned} s \cdot R_0^*(s) - 1 &= -\Gamma_{ZB1} \cdot \lambda_{ZB1} \cdot R_0^*(s) + \mu_{PZ1} \cdot Q_{ZB}^*(s) + \mu_{PZ} \cdot Q_B^*(s), \\ s \cdot Q_{ZB}^*(s) &= \Gamma_{ZB1} \cdot \lambda_{ZB1} \cdot R_0^*(s) - \mu_{PZ1} \cdot Q_{ZB}^*(s) - \Gamma_{ZB2} \cdot \lambda_{ZB2} \cdot Q_{ZB}^*(s), \\ s \cdot Q_B^*(s) &= \Gamma_{ZB2} \cdot \lambda_{ZB2} \cdot Q_{ZB}^*(s) - \mu_{PZ} \cdot Q_B^*(s). \end{aligned} \tag{7}$$

After further conversions, we acquire the symbolic (Laplace) probabilities for the ESS to stay in functional states.

$$
\begin{aligned}
R_0^*(s) &= \frac{s^2 + s\cdot\mu_{PZ} + s\cdot\mu_{PZ1} + s\cdot\Gamma_{ZB2}\cdot\lambda_{ZB2} + \mu_{PZ}\cdot\mu_{PZ1} + \mu_{PZ}\cdot\Gamma_{ZB2}\cdot\lambda_{ZB2}}{s^2\cdot\mu_{PZ} + s^2\cdot\Gamma_{ZB1}\cdot\lambda_{ZB1} + s^2\cdot\mu_{PZ1} + s^2\cdot\Gamma_{ZB2}\cdot\lambda_{ZB2} + s^3 +} \\
&\quad + s\cdot\mu_{PZ}\cdot\Gamma_{ZB1}\cdot\lambda_{ZB1} + s\cdot\mu_{PZ}\cdot\mu_{PZ1} + \\
&\quad + s\cdot\mu_{PZ}\cdot\Gamma_{ZB2}\cdot\lambda_{ZB2} + s\cdot\Gamma_{ZB1}\cdot\lambda_{ZB1}\cdot\Gamma_{ZB2}\cdot\lambda_{ZB2}, \\[4pt]
Q_{ZB}^*(s) &= \frac{s\cdot\Gamma_{ZB1}\cdot\lambda_{ZB1} + \mu_{PZ}\cdot\Gamma_{ZB1}\cdot\lambda_{ZB1}}{s^2\cdot\mu_{PZ} + s^2\cdot\Gamma_{ZB1}\cdot\lambda_{ZB1} + s^2\cdot\mu_{PZ1} + s^2\cdot\Gamma_{ZB2}\cdot\lambda_{ZB2} + s^3 +} \\
&\quad + s\cdot\mu_{PZ}\cdot\Gamma_{ZB1}\cdot\lambda_{ZB1} + s\mu_{PZ}\cdot\mu_{PZ1} + \\
&\quad + s\cdot\mu_{PZ}\cdot\Gamma_{ZB2}\cdot\lambda_{ZB2} + s\cdot\Gamma_{ZB1}\cdot\lambda_{ZB1}\cdot\Gamma_{ZB2}\cdot\lambda_{ZB2}, \\[4pt]
Q_B^*(s) &= \frac{\Gamma_{ZB1}\cdot\lambda_{ZB1}\cdot\Gamma_{ZB2}\cdot\lambda_{ZB2}}{s^2\cdot\mu_{PZ} + s^2\cdot\Gamma_{ZB1}\cdot\lambda_{ZB1} + s^2\cdot\mu_{PZ1} + s^2\cdot\Gamma_{ZB2}\cdot\lambda_{ZB2} + s^3 +} \\
&\quad + s\cdot\mu_{PZ}\cdot\Gamma_{ZB1}\cdot\lambda_{ZB1} + s\cdot\mu_{PZ}\cdot\mu_{PZ1} + \\
&\quad + s\cdot\mu_{PZ}\cdot\Gamma_{ZB2}\cdot\lambda_{ZB2} + s\cdot\Gamma_{ZB1}\cdot\lambda_{ZB1}\cdot\Gamma_{ZB2}\cdot\lambda_{ZB2},
\end{aligned}
\tag{8}
$$

The determined relationships in Equation (8) enable calculating the probabilities of an electronic security system staying in the three distinguished states, taking into account the impact of unintentionally generated electromagnetic interference. The introduction of the Γ_{ZB1} and Γ_{ZB2} coefficients of electromagnetic interference in terms of impact on the ESS enables to rationally select solutions aimed at protecting against electromagnetic interference.

5. Conclusions

The issue of operating transport telematics and electronic security systems located within railway areas is extremely important. They ensure the safety of both vehicles and passengers. For this reason, the issue related to the impact of unintentionally generated electromagnetic interference on the ESS becomes one of particular importance. They can lead to a state of partial fitness of transport telematic systems and electronic security systems, whereby the ESS is unable to execute all of its functions.

In order to determine the impact of unintentionally generated electromagnetic interference on ESS, the authors measured the low-frequency electromagnetic field generated by MV—15 and 30 kV—power lines. The obtained results enable determining numerous characteristic curves that allowed defining points of particularly high magnetic and electric field strength values. An analysis of the solution in the form of a metal housing for a container containing transport telematics and electronic security system equipment was also conducted. It was found that such a structure reduced the impact of electromagnetic interference originating from medium-voltage power lines on the analyzed systems. Therefore, it is possible to increase shielding effectiveness for a low-frequency electromagnetic field.

The conducted security system operating process analysis in terms of the impact of unintentionally generated electromagnetic interference enabled determining mathematical relationships defining the probabilities for a system to stay in distinguished states. It also allowed numerically defining the impact of applied solutions aimed at protecting against electromagnetic interference. Therefore, it is possible to rationally select solutions (structural and organizational [46]) aimed at protecting systems against electromagnetic interference.

The next stage of the study was to develop an ESS operating process model that takes into account the impact of unintentionally generated electromagnetic interference. Introducing the electromagnetic interference impact coefficient enables a rational selection of solutions aimed at protecting against electromagnetic interference. This is a significant practical aspect of the conducted test and scientific studies discussed in this article, since it has application value. As further research, the authors are planning to determine the values of electromagnetic interference impact coefficients depending on the applied design solutions.

Author Contributions: Conceptualization, K.J., J.P. and A.R.; methodology, J.P. and A.R.; validation, K.J. and J.P.; formal analysis, K.J., J.P. and A.R.; investigation, J.P. and A.R.; resources, K.J. and J.P.; data curation, K.J. and J.P.; writing—original draft preparation, K.J., J.P. and A.R.; writing—review and editing, J.P. and A.R.; visualization, K.J. and J.P.; supervision, A.R.; project administration, A.R.;

funding acquisition, J.P. and A.R. All authors have read and agreed to the published version of the manuscript.

Funding: This research received no external funding.

Institutional Review Board Statement: Not applicable.

Informed Consent Statement: Not applicable.

Data Availability Statement: The data presented in this study are available on request from the corresponding author.

Conflicts of Interest: The authors declare no conflict of interest.

References

1. Government Security Center. National Critical Infrastructure Protection Programme in Poland. Available online: https://www.gov.pl/attachment/ee334990-ec9c-42ab-ae12-477608d94eb1 (accessed on 18 August 2021).
2. Kołowrocki, K.; Soszyńska-Budny, J. Critical Infrastructure Safety Indicators. In Proceedings of the IEEE International Conference on Industrial Engineering and Engineering Management (IEEM), Bangkok, Thailand, 16–19 December 2018; pp. 1761–1764. [CrossRef]
3. Stawowy, M.; Perlicki, K.; Sumiła, M. Comparison of Uncertainty Multilevel Models to Ensure ITS Services. In *Safety and Reliability: Theory and Applications—Proceedings of the European Safety and Reliability Conference ESREL 2017, Portorož, Slovenia, 18–22 June 2017*; Cepin, M., Bris, R., Eds.; CRC Press/Balkema: London, UK, 2017; pp. 2647–2652. [CrossRef]
4. Gołębiowski, P.; Jacyna, M.; Stańczak, A. The Assessment of Energy Efficiency versus Planning of Rail Freight Traffic: A Case Study on the Example of Poland. *Energies* **2021**, *14*, 5629. [CrossRef]
5. Jacyna, M.; Żochowska, R.; Sobota, A.; Wasiak, M. Scenario Analyses of Exhaust Emissions Reduction through the Introduction of Electric Vehicles into the City. *Energies* **2021**, *14*, 2030. [CrossRef]
6. Stypułkowski, K.; Gołda, P.; Lewczuk, K.; Tomaszewska, J. Monitoring System for Railway Infrastructure Elements Based on Thermal Imaging Analysis. *Sensors* **2021**, *21*, 3819. [CrossRef] [PubMed]
7. Billinton, R.; Allan, R.N. *Reliability Evaluation of Power Systems*; Plenum Press: New York, NY, USA, 1996.
8. Klimczak, T.; Paś, J. *Basics of Exploitation of Fire Alarm Systems in Transport Facilities*; Military University of Technology: Warsaw, Poland, 2020.
9. Klimczak, T.; Paś, J. Reliability and operating analysis of transmission of alarm signals of distributed fire signaling system. *J. KONBiN* **2019**, *49*, 165–174. [CrossRef]
10. Polak, R.; Laskowski, D.; Matyszkiel, R.; Łubkowski, P.; Konieczny, Ł.; Burdzik, R. Optimizing the Data Flow in a Network Communication between Railway Nodes. In *Research Methods and Solutions to Current Transport Problems, Proceedings of the International Scientific Conference Transport of the 21st Century, Advances in Intelligent Systems and Computing, Ryn, Poland, 9–12 June 2019*; Siergiejczyk, M., Krzykowska, K., Eds.; Springer: Cham, Switzerland, 2020; Volume 1032, pp. 351–362. [CrossRef]
11. Krzykowska-Piotrowska, K.; Dudek, E.; Siergiejczyk, M.; Rosiński, A.; Wawrzyński, W. Is Secure Communication in the R2I (Robot-to-Infrastructure) Model Possible? Identification of Threats. *Energies* **2021**, *14*, 4702. [CrossRef]
12. Bednarek, M.; Dąbrowski, T.; Olchowik, W. Selected practical aspects of communication diagnosis in the industrial network. *J. KONBiN* **2019**, *49*, 383–404. [CrossRef]
13. Paś, J.; Rosiński, A.; Białek, K. A reliability-operational analysis of a track-side CCTV cabinet taking into account interference. *Bull. Pol. Acad. Sci. Tech. Sci* **2021**, *69*, e136747. [CrossRef]
14. Paś, J.; Rosiński, A.; Białek, K. A reliability-exploitation analysis of a static converter taking into account electromagnetic interference. *Transp. Telecommun.* **2021**, *22*, 217–229. [CrossRef]
15. Pas, J.; Rosinski, A.; Chrzan, M.; Bialek, K. Reliability-operational analysis of the LED lighting module including electromagnetic interference. *IEEE Trans. Electromagn. Compat.* **2020**, *62*, 2747–2758. [CrossRef]
16. RajaPriyanka, D.; ShanmugaPriya, G.; Sabashini, R.; Jacintha, V.M.E. Power generation and energy management in railway system. In Proceedings of the Third International Conference on Science Technology Engineering & Management (ICONSTEM), Chennai, India, 18 January 2017; pp. 949–954. [CrossRef]
17. Sunitha, K.; Thomas Joy, M. Effect of Soil Conditions on the Electromagnetic Field from an Impulse Radiating Antenna and on the Induced Voltage in a Buried Cable. *IEEE Trans. Electromagn. Compat.* **2018**, *61*, 990–997. [CrossRef]
18. Karami, H.; Azadifar, M.; Wang, Z.; Rubinstein, M.; Rachidi, F. Single-Sensor EMI Source Localization Using Time Reversal: An Experimental Validation. *Electronics* **2021**, *10*, 2448. [CrossRef]
19. Steczek, M.; Chudzik, P.; Szeląg, A. Application of a Non-carrier-Based Modulation for Current Harmonics Spectrum Control during Regenerative Braking of the Electric Vehicle. *Energies* **2020**, *13*, 6686. [CrossRef]
20. Geng, X.; Wen, Y.; Zhang, J.; Zhang, D. A Method to Supervise the Effect on Railway Radio Transmission of Pulsed Disturbances Based on Joint Statistical Characteristics. *Appl. Sci.* **2020**, *10*, 4814. [CrossRef]
21. Li, M.; Wen, Y.; Wang, G.; Zhang, D.; Zhang, J. A Network-Based Method to Analyze EMI Events of On-Board Signaling System in Railway. *Appl. Sci.* **2020**, *10*, 9059. [CrossRef]

22. Wróbel, Z. The Electromagnetic Compatibility in Researches of Railway Traffic Control Devices. In *Analysis and Simulation of Electrical and Computer Systems*; Springer: Cham, Switzerland, 2018; pp. 275–287. [CrossRef]
23. Smolenski, R.; Lezynski, P.; Bojarski, J.; Drozdz, W.; Long, L.C. Electromagnetic compatibility assessment in multiconverter power systems—Conducted interference issues. *Measurement* **2020**, *165*, 108119. [CrossRef]
24. Valouch, J. Technical requirements for Electromagnetic Compatibility of Alarm Systems. *Int. J. Circuits Syst. Signal Process.* **2015**, *9*, 186–191. Available online: https://www.naun.org/main/NAUN/circuitssystemssignal/2015/a522005-196.pdf (accessed on 1 October 2021).
25. Urbancokova, H.; Valouch, J.; Adamek, M. Testing of an intrusion and hold-up systems for electromagnetic susceptibility—EFT/B. *Int. J. Circuits Syst. Signal Process.* **2015**, *9*, 40–46.
26. Chmielińska, J.; Kuchta, M.; Kubacki, R.; Dras, M.; Wierny, K. Selected methods of electronic equipment protection against electromagnetic weapon. *Przegląd Elektrotechniczny* **2016**, *92*, 1–8. [CrossRef]
27. Lheurette, E. (Ed.) *Metamaterials and Wave Control*; ISTE and Wiley: London, UK, 2013.
28. Piersanti, S.; Orlandi, A.; de Paulis, F. Electromagnetic Absorbing materials design by optimization using a machine learning approach. *IEEE Trans. Electromagn. Compat.* **2018**, 1–8. [CrossRef]
29. Kozłowski, E.; Borucka, A.; Swiderski, A. Application of the logistic regression for determining transition probability matrix of operating states in the transport systems. *Eksploat. Niezawodn.–Maint. Reliab.* **2020**, *22*, 192–200. [CrossRef]
30. Hosseini, S.M.; Carli, R.; Dotoli, M. Robust optimal energy management of a residential microgrid under uncertainties on demand and renewable power generation. *IEEE Trans. Autom. Sci. Eng.* **2020**, *18*, 618–637. [CrossRef]
31. Tu, H.; Feng, H.; Srdic, S.; Lukic, S. Extreme Fast Charging of Electric Vehicles: A Technology Overview. *IEEE Trans. Transp. Electrif.* **2019**, *5*, 861–878. [CrossRef]
32. Jakubowski, J.; Kuchta, M.; Kubacki, R. D-Dot Sensor Response Improvement in the Evaluation of High-Power Microwave Pulses. *Electronics* **2021**, *10*, 123. [CrossRef]
33. Tomczuk, P.; Chrzanowicz, M.; Jaskowski, P.; Budzynski, M. Evaluation of Street Lighting Efficiency Using a Mobile Measurement System. *Energies* **2021**, *14*, 3872. [CrossRef]
34. Tomczuk, P.; Chrzanowicz, M.; Jaskowski, P. Procedure for Measuring the Luminance of Roadway Billboards and Preliminary Results. *LEUKOS* **2020**, 1–19. [CrossRef]
35. Zieja, M.; Szelmanowski, A.; Pazur, A.; Kowalczyk, G. Computer Life-Cycle Management System for Avionics Software as a Tool for Supporting the Sustainable Development of Air Transport. *Sustainability* **2021**, *13*, 1547. [CrossRef]
36. Lewczuk, K.; Kłodawski, M.; Gepner, P. Energy Consumption in a Distributional Warehouse: A Practical Case Study for Different Warehouse Technologies. *Energies* **2021**, *14*, 2709. [CrossRef]
37. Jacyna, M.; Szczepański, E.; Izdebski, M.; Jasiński, S.; Maciejewski, M. Characteristics of event recorders in Automatic Train Control systems. *Arch. Transp.* **2018**, *46*, 61–70. [CrossRef]
38. Kornaszewski, M. Modelling of exploitation process of the railway traffic control device. *WUT J. Transp. Eng.* **2019**, *124*, 53–63. [CrossRef]
39. Watral, Z.; Michalski, A. Selected Problems of Power Sources for Wireless Sensors Networks. *IEEE Instrum. Meas. Mag.* **2013**, *16*, 37–43. [CrossRef]
40. Suproniuk, M.; Skibko, Z.; Stachno, A. Diagnostics of some parameters of electricity generated in wind farms. *Przegląd Elektrotechniczny* **2019**, *95*, 105–108. [CrossRef]
41. Caban, D.; Walkowiak, T. Dependability analysis of hierarchically composed system-of-systems. In Proceedings of the Thirteenth International Conference on Dependability and Complex Systems DepCoS-RELCOMEX, Brunów, Poland, 2–6 July 2018; Springer: Cham, Switzerland, 2019; pp. 113–120. [CrossRef]
42. Duer, S. Assessment of the Operation Process of Wind Power Plant's Equipment with the Use of an Artificial Neural Network. *Energies* **2020**, *13*, 2437. [CrossRef]
43. Klimczak, T.; Paś, J. Selected issues of the reliability and operational assessment of a fire alarm system. *Eksploatacja i Niezawodn.–Maint. Reliab.* **2019**, *21*, 553–561. [CrossRef]
44. Kołowrocki, K.; Soszyńska-Budny, J. *Reliability and Safety of Complex Technical Systems and Processes: Modeling–Identification–Prediction–Optimization*; Springer: London, UK, 2011. [CrossRef]
45. Grabski, F. *Semi-Markov Processes: Applications in System Reliability and Maintenance*; Elsevier: Amsterdam, The Netherlands, 2015.
46. Suproniuk, M.; Paś, J. Analysis of electrical energy consumption in a public utility buildings. *Przegląd Elektrotechniczny* **2019**, *95*, 97–100. [CrossRef]

Article

Analysis and Assessment of Railway CCTV System Operating Reliability

Mirosław Siergiejczyk [1], Zbigniew Kasprzyk [1,*], Mariusz Rychlicki [1] and Piotr Szmigiel [2]

[1] Department of Telecommunications in Transport, Faculty of Transport, Warsaw University of Technology, Koszykowa 75 Street, 00-662 Warsaw, Poland; miroslaw.siergiejczyk@pw.edu.pl (M.S.); mariusz.rychlicki@pw.edu.pl (M.R.)

[2] PKP Polish Railway Lines S.A, Targowa 74 Street, 03-734 Warsaw, Poland; piotr.szmigiel@plk-sa.pl

* Correspondence: zbigniew.kasprzyk@pw.edu.pl

Abstract: The article reviews the history and the direction of development for railway CCTV (Closed-Circuit TeleVision) systems. The authors described the CCTV system at PKP Polskie Linie Kolejowe S.A. and the associated network and server infrastructure. The authors proposed an operational model for a centralized CCTV system that assumes states of partial fitness, in accordance with the regulations of the national railway infrastructure administrator. The aim of the paper is to review, analyse, and evaluate the operational reliability of railroad video monitoring systems in relation to the assumptions of the national railroad infrastructure manager using an operational model. A unified system structure is presented in the article. The model was used as a base to calculate the probabilities for the system while staying in the assumed states. Calculations showed that a centralized CCTV system is characterized by high reliability and satisfies the expectations of PKP Polskie Linie Kolejowe S.A. in this respect. The obtained result of 99.88% probability of leaving the analysed video surveillance system in a fully operational condition within a year indicates a high level of security of the applied solutions in such a large system. The analysed system is one of the largest such solutions designed in the European Union and the largest in Poland, which is an important contribution to the development and implementation of such extensive video surveillance systems in the future. The research question is whether the extensive centralized railway CCTV systems will meet the requirements of PKP Polskie Linie Kolejowe S.A.

Keywords: video surveillance; railway safety systems; rail safety systems

Citation: Siergiejczyk, M.; Kasprzyk, Z.; Rychlicki, M.; Szmigiel, P. Analysis and Assessment of Railway CCTV System Operating Reliability. *Energies* **2022**, *15*, 1701. https://doi.org/10.3390/en15051701

Academic Editors: Stanisł Duer and Ahmadieh Khanesar Mojtaba

Received: 9 January 2022
Accepted: 22 February 2022
Published: 24 February 2022

Publisher's Note: MDPI stays neutral with regard to jurisdictional claims in published maps and institutional affiliations.

1. Introduction and Analysis of the Issue

Until 2017, only 170 stations and stops out of 2563 active facilities managed by PKP Polskie Linie Kolejowe S.A. (Polish railway infrastructure administrator) had site CCTV, with as many as 100 being obsolete analogue systems. The company lacked regulations on CCTV systems dedicated to passenger infrastructure, which is why newly built systems were non-uniform and/or functionally poor. Only island-type systems were constructed, without the possibility for remote operation. Viewing stations were located typically in Local Control Centres (LCC) or command signal boxes, with viewing conducted locally and occasionally by employees dealing with railway traffic management, which resulted in a low effectiveness relative to the assumed objectives, i.e., protection of people and property.

The previous lack of widespread application of CCTV systems contributed to the low feeling of passenger safety. In a 2017 survey that covered 800 passengers, the main threats indicated in relation to train stations were banditry and the unpredictable behaviour of intoxicated or mentally unstable people [1] (Figure 1). The national infrastructure administrator was approached with questions about surveillance systems—both by passengers and local government authorities. In addition, the existing scientific studies indicated the significant role of CCTV systems in preventing crime at local railway facilities [2].

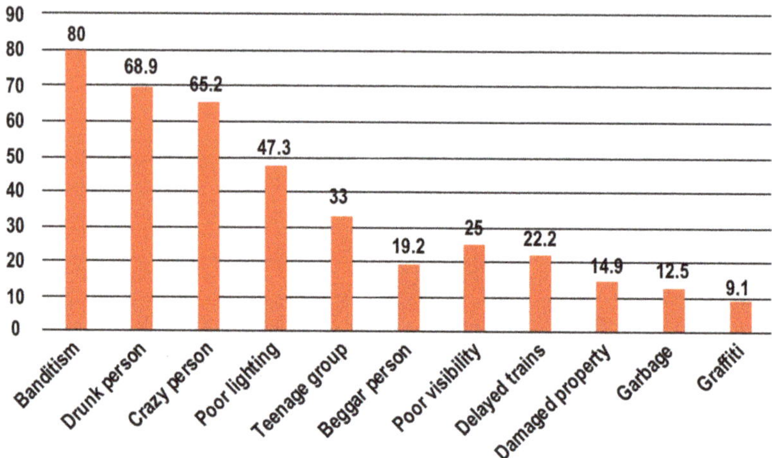

Figure 1. The causes for fear among Polish passengers at railway stations. Source: own study based on [1].

Trying to meet the safety-related expectations of the passengers halfway, as well as in recognition of the threat associated with terrorist attacks in Europe, the national railway infrastructure administrator in Poland decided to respond through developing and implementing a centralized CCTV system that would ultimately cover all the largest passenger stations and stops in Poland. The Ipi-4 instruction entitled "Guidelines for designing and constructing Closed-Circuit Television (CCTV) systems at passenger handling facilities" has been in force at the Polish railway since 2017. It sets out the principles in terms of designing, constructing and operating a modern CCTV system within passenger infrastructure areas at railway stations. The first CCTV systems based on these instructions were commissioned in 2020, upon the commissioning of the modernized railway line No. 447, and the following stations were covered by a surveillance system satisfying the requirements [3]: Warszawa Włochy, Warszawa Ursus, Warszawa Ursus Niedźwiadek, Piastów, Pruszków, Parzniew, Brwinów and Milanówek. The number of train stations with installed modern surveillance systems has been systematically growing ever since.

The next step of PKP Polskie Linie Kolejowe S.A. was announcing a tender to the execution of central system parts—a server room with a planned integration platform that constitutes a key element of the system, as well as a supervision centre that will monitor all railway stations equipped with a surveillance system in line with [3] 24 h a day. A contract for the execution of this task was concluded in 2020—the selected contractor is the Spanish company Aldesa. It is anticipated that the centralized surveillance system will cover 200 railway stations by 2023, with a target of approximately 1050 stations.

There are numerous solutions in terms of video surveillance in the world; however, most of them are not directly associated with railways, and the CCTV structures are significantly different than the one described in this paper.

The authors of [4] believe that video recordings collected and processed in CCTV systems contain important personal data, the reliable protection of which is one of the key operating parameters. Given the fact that the development of artificial technologies in the coming years will probably contribute to an equally dynamic development of intelligent video analysis, decisive measures have to be taken, aimed at protecting such data. Traditional video data protection methods involve masking or simple encryption and do not offer efficient and safe CCTV video search algorithms that are based on video metadata. Moreover, such data are usually stored in the form of plain text. Based on these premises, the authors of this article propose a COP transformation technique, which has the advantage of significantly increasing the efficiency and safety of video metadata. This is possible owing to the fact that a query is sent to a database in the same manner as for text files, not leaving plaintext within the processed files. The authors indicated that in the course of creating queries for searching metadata, the data that utilize COP transformation offer higher query processing efficiency versus traditional data utilizing text files. In other words, databases implemented within COP transformation may not only execute matching and range queries, but also queries that utilize join-based for multiple base tables. Moreover, they can create simple queries during a statistical analysis of meta-information.

The publication in [5] is also noteworthy. The implementation and provision of communication between vehicles through a wide range of additional application and services have significant impacts on their operation, from safety on roads for wheeled and rail vehicles, to supervision and management of traffic, and even infotainment. However, each application imposes its own limitations regarding the quality of service (QoS) on information exchange. The required efficiency of offered services significantly differs in terms of bandwidth, latency and communication reliability. For example, high-bandwidth applications, such as video streaming, require highly reliable communication. However, damping of an IEEE 802.11p/DSRC communication link caused by static and mobile blocking objects deteriorates link quality and may threaten QoS requirements of supported apps. In contrast, hybrid architecture with two interfaces may offer an emergency switching or backup route creation mechanism and be used for occasionally offloading transmission through more reliable links, such as cellular networks.

The authors of this article propose an approach towards hybrid communication that is based on 4G/LTE and IEEE 802.11p technologies in order to support V2X video streaming applications. The authors conducted extensive studies based on measurements using a field station configuration with a software protocol stack. Field results were collected under various network conditions and in the presence of various blocking objects (LOS, NLOS_V and NLOS_B). The results show that the proposed solution is practically feasible and offers a significant increase in communication reliability. It also enables one to expand the reliable communication range. Furthermore, smooth network switching owing to RAT selection taking into account QoS and the VHO algorithm enables trouble-free and reliable video streaming without failures and interruptions. The proposed approach is a manifestation of an effective compromise between using the IEEE 802.11p/DSCR interface and ensuring a better-quality video streaming service.

Paper [6] presented a traditional CCTV system with marked operational limitations, mainly due to a fixed and preset surveillance pattern. This may reduce system reliability and cause increased generation of false alarms, which translates to increased system processing activity, leading to increased consumption of system resources and energy. In their work, the authors suggested improving these CCTV system operating parameters through a smart combination of a sensor assembly with two cameras, actuators and a lighting module, as well as an implementation of economic built-in processors. The key to success was keeping most CCTV system elements on standby. An exception was made only for system sensors with low power consumption. An effective combination of a sensor assembly with a developed classifier enables one to reduce the generation of a false alarm and improve the reliability of the entire system. In addition, the result was a reduced use of system memory and energy consumption, as well as transmission link capacity, compared to traditional counterparts, which significantly contributed to improved operating parameters of the analysed video surveillance system parameters.

The authors of [7] presented an issue associated with data encryption processes that are a significant burden for the efficiency of modern CCTV systems. These processes are simultaneously the basic protection mechanism for collected and processed data, which is an extremely important issue from the perspective of the reliability and operation of such systems. The hardware support for this process may considerably impact the improvement of these parameters. The authors discussed a systematic real-time video data encryption and decryption methodology based on the idea of chaos in terms of system engineering and analysing data processing algorithms. The proposed system design and the conducted algorithm analyses were verified using an FPGA hardware platform, employing the Verilog hardware description language (Verilog HDL). The results of the analysis conducted using TETU01 statistical test sets and a differential analysis confirmed increased efficiency and reliability of such solutions relative to CCTV systems.

In [8], the authors proposed a method for improving image quality through the application of data matching with the histogram shaping technique. Image quality is one of the basic parameters of video surveillance systems. It is of particular importance in relation to the systems operated by the railway, where reliability and safety aspects are especially emphasized. Experiments were conducted using 1280×780 px images collected from a CCTV System. The .jpg image format in the form of RGB42 and an .avi video file were selected for analysis. The results of the experiment conducted within a railway station CCTV system indicated satisfactory efficiency of the utilized histogram shaping method for image detection under good and satisfactory lighting conditions, significantly contributing to improving the operating parameters of the system.

The authors of [9] present an underground railway communication system with installed DDS data exchange intermediary software that offers QoS at a preset reliability level. This can be a solution that enables one to maintain information traffic within the network. In such applications as a DDS system for monitoring passenger traffic, it can store the data captured from video file via CCTV cameras distributed within the network, regardless of the encountered communication interruptions. In the phase of contact between stations, data exchanged between train carriage and railway stations upon a stopping train may still be stored, even if the train is disconnected from global networks. This is possible owing to the fact that the system is based on an architecture that comprises more than two domains referenced to broader general architecture, with each of the domains representing an element of an underground railway station. The first domain is the one inside the carriage. The second is located at the train station platform. Further ones are locations at upper station floors. The study involved a simulation aimed at testing the exchange of system video data with two domains, through changing the number of subscribers and the size of sent messages. The results showed that adding a larger number of subscribers did not impact system efficiency. Owing to this, in order to improve the reliability parameters, one can conveniently implement such a system for continuous transmission and collection of centralized information.

In [10], the authors presented an approach to the CCTV analysis of the crowd and of human behaviour in dangerous situations at railway stations. Railway stations are public places that experience significant concentrations of people. For this reason, they are facilities highly exposed to the hazard of crowd panic that can result from a catastrophe or a terrorist attack. Crowd analysis is essential from the perspective of security and surveillance, as it detects abnormal behaviour, thus contributing to reducing the risk of an accident. However, most classic approaches to controlling overload and congestion are based on a hardware approach. This article proposed an approach based on software and an early warning system (CCEWS) for controlling overload through the object detection and tracking technique. Object detection was conducted in accordance with the faster R-CNN architecture. The results revealed that the faster R-CNN model, together with the Google inception resnet v2 CNN model, provided significantly better outcomes, detecting people in appropriate frames of a CCTV image with an accuracy of 93.503% at a 28 FPS. Therefore, it was demonstrated that the general human detection accuracy and FPS result for the Faster R-CNN were significantly better than in the case of other methods. It was also shown that it was possible, under normal CCTV operating conditions, for alarm signals to be generated for each frame with a higher possibility of unfortunate event potential, such as panic due to crowd traffic congestion within a railway facility.

The authors of [11] discuss the issue of the accumulation of a large number of people at railway stations—hence potential targets of terrorist attacks. Access to railway stations is open and is subject to less control than, e.g., airports. For this reason, the reliability of CCTV systems is the key element in terms of improving security at such facilities. One of the fundamental aspects is efficient, rapid and reliable detection of left objects, such as abandoned luggage. The proposed algorithm can detect lighting changes and adapt to them, which enables one to avoid the generation of false stationary objects associated with poor efficiency. This significantly impacts CCTV system reliability.

Paper [12] presented an issue related to threats within railway facilities that arise from intrusion and property theft, often of considerable value. In order to prevent them, protection systems based on video surveillance are widely used in modern railway systems. It uses an adaptive feature distribution extractor to segment railway tracks through the complete utilization of strong, linear railway scene characteristics and typical categories of local monitored areas. Owing to the application of the presented algorithm, a railway intrusion detection system can automatically and accurately define the boundaries of a monitored scene in real-time and significantly improve its operating efficiency.

The authors of [13] also discuss the issue of threats associated with physical assault and banditry, which are some of the main concerns of rail transport passengers. In order to eliminate such behaviour, it is crucial to introduce reliable and highly efficient CCTV systems. However, their popularization and expansion lead to a situation in which their operators are not able to process the huge amount of information that such systems continuously provide. For this reason, efficient operation of such systems requires the implementation of efficient algorithms supporting automatic and reliable detection of specific situations, such as violence against passengers. In their paper, the authors proposed a three-tier, comprehensive violence detection framework based on deep learning. The experimental results obtained in the course of the studies on various comparative data sets confirm that the proposed method is best suited for detecting violence within CCTV systems and leads to improving their operating parameters, since it achieves higher accuracy than a number of techniques currently applied in this respect. The method is thus efficient that the authors intend to ensure its implementation on devices with limited resources, which will enable its effective deployment even on IoT devices.

Paper [14], which presents the issue of crowd analysis, is also noteworthy. Paper [14] describes the development and evaluation of a multi-stream, convolutional neural network that receives image as input data and generates a density map that comprehensively shows the spatial distribution of people. In order to solve the assumed problems associated with crowd counting, such as extremely unlimited scale and perspective changes, the network architecture utilizes vulnerable fields of various size for each stream. Furthermore, the impact of the two most common trends on generating truths is tested, and a hybrid method based on detecting small faces and scale interpolation is proposed. The experiments conducted on two data sets, UCF-CC-50 and ShanghaiTech, demonstrate that the application of basic truth generation methods enables one to obtain excellent results.

Another study was described in [15]. The paper proposes a smart video surveillance system for level crossings. In this case, a smart video surveillance system starts with extraction, detection and tracking objects moving within a level crossing area (threat area), using the proposed variance-based method. This new method is based on subtracting five background frames, differentiating five frames and calculating variance in order to detect and track objects. The variance-based method involves calculating the variability of columns and rows in video frames, where image pixel intensity changes determine the position of a moving object and is used to locate and track objects in a video. This algorithm enables accurate detection of an object within a hazardous area, with minimum calculation times.

Paper [16] presents issues related to the impact of electromagnetic interference on track-side cabinets of closed-circuit television system operating in the railroad transport environment. The paper develops an operational model including electromagnetic interference. The presented results allow for a numerical evaluation of different types of solutions that can be used to mitigate the impact of electromagnetic interference on the functioning of the system.

Paper [17] presents a study on traffic analysis using computer vision techniques. The traffic volume analysis is based on a CCTV system. This method is challenging in the perspective when there are many vehicle traffic streams at the intersection. In this paper, research is conducted on the processes of improving CCTV-based vehicle counting for traffic analysis. In particular, a comprehensive framework with multiple classes and movements for vehicle counting has been proposed. This paper presents deep learning methods for vehicle detection and tracking. A suitable trajectory approach for monitoring vehicle movements using highlighted region tracking to improve counting performance is presented.

Paper [18] proposes a video-based smoke detection technique for early warning in fire surveillance systems. The paper presents an algorithm to detect smoke in a limited video surveillance environment, both indoors and outdoors. The proposed method uses Kalman estimator, colour analysis, image segmentation, speckle labelling, geometric feature analysis, and M of N decoders to extract the alarm signal at a well-defined time. The proposed smoke detection technique is flexible in terms of input camera type, size, and frame rate and has been implemented on a low-cost platform accessible through a web browser.

2. PKP Polskie Linie Kolejowe S.A. CCTV System Architecture

The central part of the planned CCTV system of PKP Polskie Linie Kolejowe S.A. is the Main Server Room (GS), which houses central and executive system elements. The most important element is the integrated platform called PSIM (Physical Security Information Management), which is an application installed as a virtual instance on physical servers with internal redundancy (Figure 2). The integration platform acts as an intermediary between all system devices. Using a protocol appropriate for a given device, and through access to an Application Programming Interface (API), the PSIM is able to ensure communication between devices, regardless of the ICT standards of such devices. In other words, even in the case of using devices that operate solely based on protocols native for a given manufacturer, it enables the system to operate as a uniform whole, ensuring full control over all processes. PSIM ensures a uniform graphical user interface, regardless of the integrated solution, which is why operators do not have to learn the operation of each added system. This is particularly important in the case of operating and management systems on video recording devices (VMS—Video Management System). These systems significantly differ in terms of interface and are designed based on various management methods. A VMS administrator guidebook can be several hundred pages long, and the same functions may be called and implemented differently, depending on the manufacturer. System operators and administrators at CBIP (Passenger Infrastructure Security Centre) have PSIM-access to all systems through a PSIM client application installed on workstations.

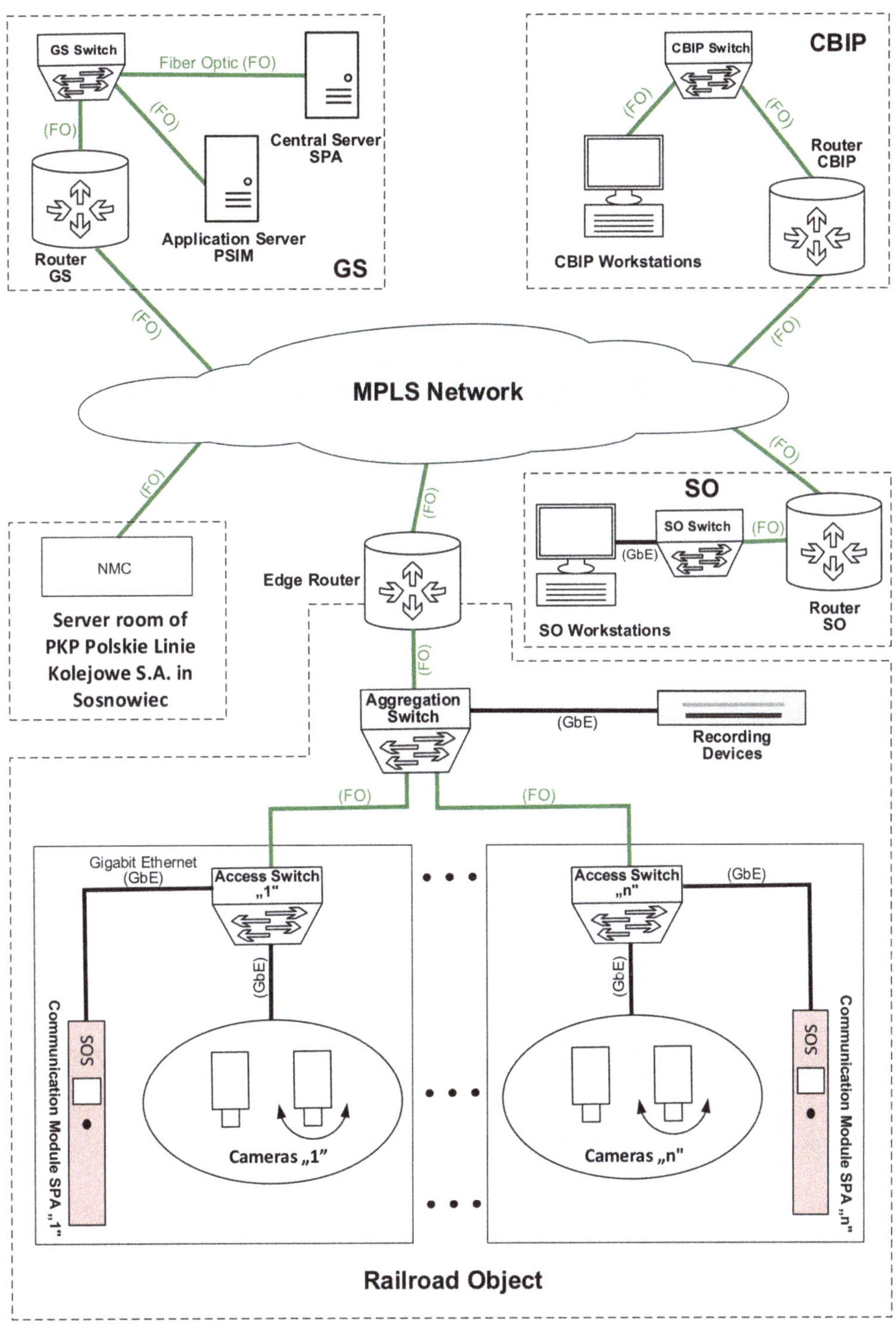

Figure 2. CCTV architecture. Source: own study based on [19–21].

The following systems are subject to integration:

- Facility CCTV—through VMS integration, less often directly through integrating individual cameras through the ONVIF protocol—an open-source protocol that ensures a uniform camera management and communication standard.
- CAS—through integration of a central server located in the GS—a VoIP (Voice over IP) switchboard. The integration is achieved only through ensuring a uniform system operation interface, but also enables functionally linking CAS and CCTV, e.g., through the CBIP-displaying image from a camera installed nearest to the SPA communication module upon an emergency call.
- ACS—Access Control System located in ICT racks. CBIP operators receive information on employee access to the racks.
- IDS—Intrusion Detection system located in ICT racks. Racks are equipped with door opening sensors, as well as vibration sensors, which is why not only unauthorized rack opening is signalled, but also, e.g., attempts to overturn it.

PSIM also ensures the integration of peripherals (air conditioning units in ICT racks, managed power supply strips—CBIP operators are able to monitor their condition). The GS also houses authentication servers (conducted with the use of a device-specific protocol, e.g., RADIUS—Remote Authentication Dial In User Service) and update servers (their role is to update software and the microcode of all PSIM-integrated devices, executing test and collective updates, and the potential restoration of previous versions using an update schedule, e.g., several devices are ongoing test updates; after verifying their efficiency, the updates are executed in packages of several dozen devices, in order not to make the entire system inefficient for the duration of a software/microcode update).

All devices in the GS utilize a common data repository, i.e., drive matrices equipped with an appropriate number of HDD SAS (Serial Attached SCSI) hard drives. Site CCTV is installed at individual railway stations. The system consists of an assembly of executive devices (cameras, infrared radiators, train start/end sensors) as well as CAS communication modules. Camera video streams are sent to an assigned recording device.

A native resolution image is saved on a recording device drive and is available on demand for CBIP operators—such a solution has been adopted to reduce network load since a single 4K camera is able to generate 20 Mbit, even when using the most up-to-date H.265+ compression. One or several stations are monitored live at Viewing Stations (VS) located in their direct vicinity and equipped with a single workstation (identical to CBIP) vs. being usually located in railway utility buildings and operated typically by Railway Security (SOK) employees. They do not allow for administrative access—they can be managed only from the CBIP. From the network perspective, each station has a LAN network, the access switches of which supply actuators through PoE (Power over Ethernet). These switches, as well as the recording devices, are coupled to an aggregation switch cooperating with a site router (demarcation point)—this is the point of contact with the WAN network, where site routers connect with WAN edge devices. An access network diagram is shown in Figure 3.

The IP MPLS backbone network owned by PKP Polskie Linie Kolejowe S.A. is the WAN. This network is to ultimately cover the entire country and consists of thousands of kilometres of fibre optic cables. It is based on a multiple-ring topology—it has three router layers (access, aggregation and backbone). The topology has been selected in order to ensure high accessibility, important in terms of network purpose—it is not only used for CCTV or CAS, but also for railway communication and control systems, e.g., GSM-R and ETCS. A teletransmission network diagram is shown in Figure 4. Network elements are managed at the network management centre in Sosnowiec and, complementary, at CBIP. The network is managed by PKP Polskie Linie Kolejowe S.A.

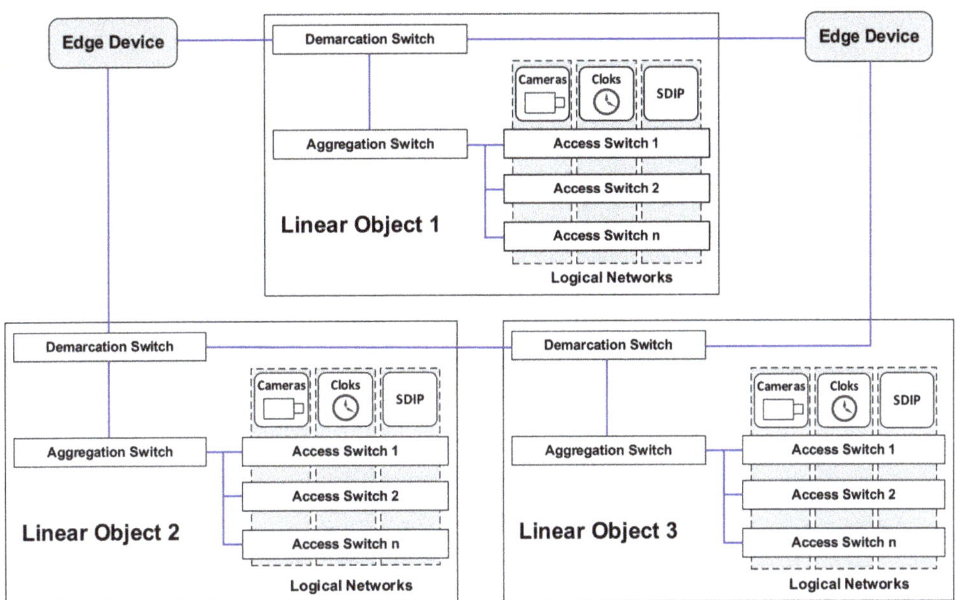

Figure 3. Diagram of a line railway facility access network. Source: own study based on [22].

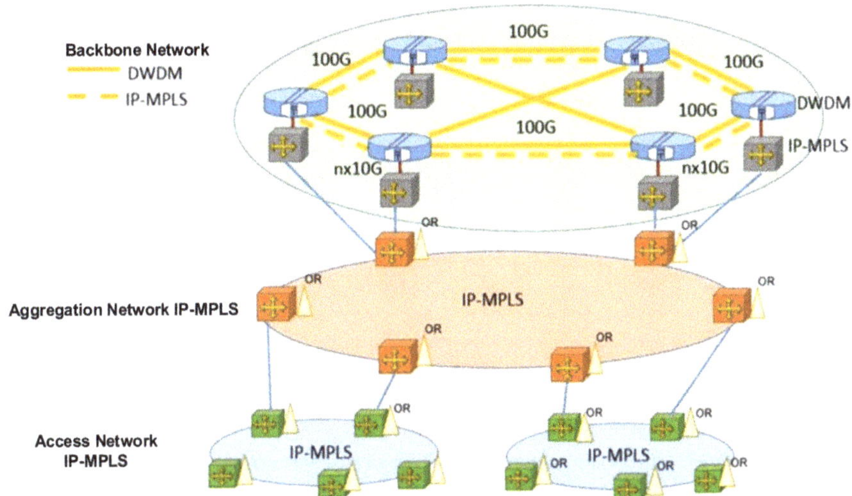

Figure 4. PKP Polskie Linie Kolejowe S.A. teletransmissions network diagram. Source: own study based on [22].

3. PKP Polskie Linie Kolejowe S.A. CCTV System Operating Model

The reliability model was developed in accordance with generally accepted reliability analysis of telecommunication networks using Markov models [18]. The following assumptions were adopted for the development of a PKP Polskie Linie Kolejowe S.A. closed-circuit television system:

- The system is in the state of full fitness (S_1) if and only if its objective function is fulfilled, i.e., assumptions from [19]: 100% of the passenger infrastructure area at each railway facility is covered by monitoring, with mapping detail appropriate for the category (class) of the stop/station where the system is installed—expressed in pixels per metre.

- Damage to a single camera will mean the creation of a dead zone (facility coverage will drop below 100%) or a zone with reduced mapping detail relative to the requirements [19]. Therefore, the system will cease to satisfy the requirements; however, this will apply only to a certain area of a single railway facility. The system can essentially be considered operable, however not exhibiting the expected parameters (S_2). The system will not transition into state S_2 due to short-term signal losses—cameras are equipped with SD cards that the video stream is sent to when it is impossible to establish a connection with the recording device—the system will then remain in the state of fitness S_1;
- Damage to one of the access switches (that ensure not only data transmission, but also power for cameras via the Power over Ethernet—PoE) will result in a lack of monitoring coverage of the entire area, e.g., a given station platform. The switches serve independent areas of a railway facility. Access switches are not ensured by any kind of hardware redundancy. The system ceases to fulfil its task for one of the areas; however, it will still be possible to establish alarm communication with other areas of the facility (e.g., underground passages, other platforms). Therefore, the system will transition to the state of partial fitness S_3.
- An event that takes an entire service out of operation is damage to the aggregation switch, a pair of aggregation routers (fibre optic loop start and end for a certain line segment) or a pair of backbone network edge routers. This will make video monitoring and alarm communication impossible for one or numerous railway facilities, while resulting in a transition into the state of partial system unfitness S_4.
- Due to no GS georedundancy, any event that leads to the unfitness of key system elements (e.g., PSIM hardware platform) results in the full unfitness of the entire system—S_5—since a damage to server room elements causes the system to stop working. The PSIM platform has been constructed with local hardware redundancy, which is why in the event of hardware platform failures, the S_5 state may be considered only in the case of a simultaneous damage to both physical PSIM servers or a damage to the redundant pair of switches/routers in the GS.
- Repair times result from the assumptions adopted in [19–21] and are implemented by the hardware vendor and by a service contractor under a maintenance contract, after the warranty period expires.
- Transitions from less severe failure states to more severe are possible; however, repair in such an event is simultaneous for all damaged devices—for logistic reasons. Therefore, there is no probability of a transition from a state of a more severe failure to a state of less severe system failure. The operational model is shown in Figure 5.

The designations in Figure 5 show the following system functions and transition intensities:

- $R_0(t)$—probability function for a CCTV system in the state of full fitness S_1;
- $Q_2(t)$—probability function for a CCTV system in the state of partial fitness S_2;
- $Q_3(t)$—probability function for a CCTV system in the state of partial fitness S_3;
- $Q_4(t)$—probability function for a CCTV system in the state of partial unfitness S_4;
- $Q_5(t)$—probability function for a CCTV system in the state of full unfitness S_5;
- λ_{12}—intensity of transitions from the state of full fitness S_1 to the state of partial fitness S_2;
- μ_{21}—intensity of transitions from the state of partial fitness S_2 to the state of full fitness S_1;
- λ_{13}—intensity of transitions from the state of full fitness S_1 to the state of partial fitness S_3;
- μ_{31}—intensity of transitions from the state of partial fitness S_3 to the state of full fitness S_1;

- λ_{14}—intensity of transitions from the state of full fitness S_1 to the state of partial unfitness S_4;
- μ_{41}—intensity of transitions from the state of partial unfitness S_4 to the state of full fitness S_1;
- λ_{15}—intensity of transitions from the state of full fitness S_1 to the state of full unfitness S_5;
- μ_{51}—intensity of transitions from the state of full unfitness S_5 to the state of full fitness S_1;
- λ_{23}—intensity of transitions from the state of partial fitness S_2 to the state of partial fitness S_3;
- λ_{34}—intensity of transitions from the state of partial fitness S_3 to the state of partial unfitness S_4;
- λ_{45}—intensity of transitions from the state of partial unfitness S_4 to the state of full unfitness S_5.

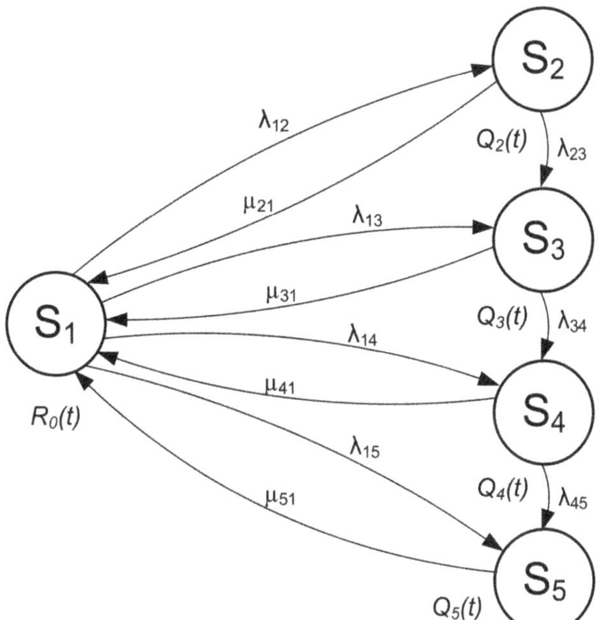

Figure 5. Operational model of a CCTV system. Source: own study.

The operating model in Figure 5 is shown in the form of Kolmogorov–Chapman equations.

$$R'_0(t) = -\lambda_{12}\cdot R_0(t) + \mu_{21}\cdot Q_2(t) - \lambda_{13}\cdot R_0(t)$$
$$+\mu_{31}\cdot Q_3(t) - \lambda_{14}\cdot R_0(t) + \mu_{41}\cdot Q_4(t) - \lambda_{15}\cdot R_0(t) + \mu_{51}\cdot Q_5(t) \tag{1}$$

$$Q'_2(t) = \lambda_{12}\cdot R_0(t) - \mu_{21}\cdot Q_2(t) - \lambda_{23}\cdot Q_2(t) \tag{2}$$

$$Q'_3(t) = \lambda_{13}\cdot R_0(t) - \mu_{31}\cdot Q_3(t) - \lambda_{34}\cdot Q_3(t) + \lambda_{23}\cdot Q_2(t) \tag{3}$$

$$Q'_4(t) = \lambda_{14}\cdot R_0(t) - \mu_{41}\cdot Q_4(t) - \lambda_{45}\cdot Q_4(t) + \lambda_{34}\cdot Q_3(t) \tag{4}$$

$$Q'_5(t) = \lambda_{15}\cdot R_0(t) - \mu_{51}\cdot Q_5(t) + \lambda_{45}\cdot Q_4(t) \tag{5}$$

Assuming baseline conditions:

$$R_0(0) = 1 \tag{6}$$

$$Q_2(0) = Q_3(0) = Q_4(0) = Q_5(0) = 0 \tag{7}$$

Applying the Laplace transform, the following system of linear equations is obtained:

$$s \cdot R^*_0(s) - 1 = -\lambda_{12} \cdot R^*_0(s) + \mu_{21} \cdot Q^*_2(s) - \lambda_{13} \cdot R^*_0(s) + \mu_{31} \cdot Q^*_3(s) - \lambda_{14} \cdot R^*_0(s)$$
$$+ \mu_{41} \cdot Q^*_4(s) - \lambda_{15} \cdot R^*_0(s) + \mu_{51} \cdot Q^*_5(s) \tag{8}$$

$$s \cdot Q^*_2(s) = \lambda_{12} \cdot R^*_0(s) - \mu_{21} \cdot Q^*_2(s) - \lambda_{23} \cdot Q^*_2(s) \tag{9}$$

$$s \cdot Q^*_3(s) = \lambda_{13} \cdot R^*_0(s) - \mu_{31} \cdot Q^*_3(s) - \lambda_{34} \cdot Q^*_3(s) + \lambda_{23} \cdot Q^*_2(s) \tag{10}$$

$$s \cdot Q^*_4(s) = \lambda_{14} \cdot R^*_0(s) - \mu_{41} \cdot Q^*_4(s) - \lambda_{45} \cdot Q^*_4(s) + \lambda_{34} \cdot Q^*_3(s) \tag{11}$$

$$s \cdot Q^*_5(s) = \lambda_{15} \cdot R^*_0(s) - \mu_{51} \cdot Q^*_5(s) + \lambda_{45} \cdot Q^*_4(s) \tag{12}$$

The probabilities for a video surveillance system to stay in given operating states of the system are as follows:

$$R^*_0(s) = \frac{(\lambda_{23} + s + \mu_{21}) \cdot (\lambda_{34} + s + \mu_{31}) \cdot (\lambda_{45} + s + \mu_{41}) \cdot (s + \mu_{51})}{\begin{array}{c}(\lambda_{13} \cdot (\lambda_{23} \cdot \mu_{31} \cdot (\lambda_{45} + s + \mu_{41}) \cdot (s + \mu_{51}) + \mu_{21} \cdot (\lambda_{34} + s + \mu_{31}) \cdot (\lambda_{45} + s + \mu_{41}) \cdot (s + \mu_{51}) \\ + \lambda_{23} \cdot \lambda_{34} \cdot (s \cdot \mu_{41} + (\lambda_{45} + \mu_{41}) \cdot \mu_{51})) + (\lambda_{23} + s + \mu_{21}) \cdot (\lambda_{13} \\ \cdot (\lambda_{34} \cdot \lambda_{45} \cdot \mu_{51} + \lambda_{34} \cdot \mu_{41} \cdot (s + \mu_{51}) \\ + \mu_{31} \cdot (\lambda_{45} + s + \mu_{41}) \cdot (s + \mu_{51})) + (\lambda_{34} + s + \mu_{31}) \cdot (\lambda_{14} \cdot (s \cdot \mu_{41} + (\lambda_{45} + \mu_{41}) \cdot \mu_{51}) \\ + (\lambda_{45} + s + \mu_{41}) \cdot (\lambda_{15} \cdot \mu_{51} - (\lambda_{12} + \lambda_{13} + \lambda_{14} + \lambda_{15} + s) \cdot (s + \mu_{51})))))\end{array}} \tag{13}$$

$$Q^*_2(s) = \frac{(\lambda_{12} \cdot (\lambda_{34} + s + \mu_{31}) \cdot (\lambda_{45} + s + \mu_{41}) \cdot (s + \mu_{51}))}{\begin{array}{c}(s \cdot (\lambda_{12} \cdot (\lambda_{23} \cdot (\lambda_{45} + s + \mu_{41}) \cdot (s + \mu_{51}) + (\lambda_{34} + s + \mu_{31}) \cdot (\lambda_{45} + s + \mu_{41}) \cdot (s + \mu_{51}) \\ + \lambda_{23} \cdot \lambda_{34} \cdot (\lambda_{45} + s + \mu_{51})) + (\lambda_{23} + s + \mu_{21}) \cdot ((\lambda_{34} + s + \mu_{31}) \cdot ((\lambda_{45} + s + \mu_{41}) \cdot (\lambda_{15} + s + \mu_{51}) \\ + \lambda_{14} \cdot (\lambda_{45} + s + \mu_{51})) + \lambda_{13} \cdot ((\lambda_{45} + s + \mu_{41}) \cdot (s + \mu_{51}) + \lambda_{34} \cdot (\lambda_{45} + s + \mu_{51})))))\end{array}} \tag{14}$$

$$Q^*_3(s) = \frac{((\lambda_{12} \cdot \lambda_{23} + \lambda_{13} \cdot (\lambda_{23} + s + \mu_{21})) \cdot (\lambda_{45} + s + \mu_{41}) \cdot (s + \mu_{51}))}{\begin{array}{c}(s \cdot (\lambda_{12} \cdot (\lambda_{23} \cdot (\lambda_{45} + s + \mu_{41}) \cdot (s + \mu_{51}) + (\lambda_{34} + s + \mu_{31}) \cdot (\lambda_{45} + s + \mu_{41}) \cdot (s + \mu_{51}) \\ + \lambda_{23} \cdot \lambda_{34} \cdot (\lambda_{45} + s + \mu_{51})) + (\lambda_{23} + s + \mu_{21}) \cdot ((\lambda_{34} + s + \mu_{31}) \cdot ((\lambda_{45} + s + \mu_{41}) \cdot (\lambda_{15} + s + \mu_{51}) \\ + \lambda_{14} \cdot (\lambda_{45} + s + \mu_{51})) + \lambda_{13} \cdot ((\lambda_{45} + s + \mu_{41}) \cdot (s + \mu_{51}) + \lambda_{34} \cdot (\lambda_{45} + s + \mu_{51})))))\end{array}} \tag{15}$$

$$Q^*_4(s) = \frac{((\lambda_{12} \cdot \lambda_{23} \cdot \lambda_{34} + (\lambda_{23} + s + \mu_{21}) \cdot (\lambda_{13} \cdot \lambda_{34} + \lambda_{14} \cdot (\lambda_{34} + s + \mu_{31}))) \cdot (s + \mu_{51}))}{\begin{array}{c}(s \cdot (\lambda_{12} \cdot (\lambda_{23} \cdot (\lambda_{45} + s + \mu_{41}) \cdot (s + \mu_{51}) + (\lambda_{34} + s + \mu_{31}) \cdot (\lambda_{45} + s + \mu_{41}) \cdot (s + \mu_{51}) \\ + \lambda_{23} \cdot \lambda_{34} \cdot (\lambda_{45} + s + \mu_{51})) + (\lambda_{23} + s + \mu_{21}) \cdot ((\lambda_{34} + s + \mu_{31}) \cdot ((\lambda_{45} + s + \mu_{41}) \cdot (\lambda_{15} + s + \mu_{51}) \\ + \lambda_{14} \cdot (\lambda_{45} + s + \mu_{51})) + \lambda_{13} \cdot ((\lambda_{45} + s + \mu_{41}) \cdot (s + \mu_{51}) + \lambda_{34} \cdot (\lambda_{45} + s + \mu_{51})))))\end{array}} \tag{16}$$

$$Q^*_5(s) = \frac{(\lambda_{12} \cdot \lambda_{23} \cdot \lambda_{34} \cdot \lambda_{45} + (\lambda_{23} + s + \mu_{21}) \cdot (\lambda_{13} \cdot \lambda_{34} \cdot \lambda_{45} + (\lambda_{34} + s + \mu_{31}) \cdot (\lambda_{14} \cdot \lambda_{45} + \lambda_{15} \cdot (\lambda_{45} + s + \mu_{41}))))}{\begin{array}{c}(s \cdot (\lambda_{12} \cdot (\lambda_{23} \cdot (\lambda_{45} + s + \mu_{41}) \cdot (s + \mu_{51}) + (\lambda_{34} + s + \mu_{31}) \cdot (\lambda_{45} + s + \mu_{41}) \cdot (s + \mu_{51}) \\ + \lambda_{23} \cdot \lambda_{34} \cdot (\lambda_{45} + s + \mu_{51})) + (\lambda_{23} + s + \mu_{21}) \cdot ((\lambda_{34} + s + \mu_{31}) \cdot ((\lambda_{45} + s + \mu_{41}) \cdot (\lambda_{15} + s + \mu_{51}) \\ + \lambda_{14} \cdot (\lambda_{45} + s + \mu_{51})) + \lambda_{13} \cdot ((\lambda_{45} + s + \mu_{41}) \cdot (s + \mu_{51}) + \lambda_{34} \cdot (\lambda_{45} + s + \mu_{51})))))\end{array}} \tag{17}$$

A computer simulation enables one to quickly determine the impact of changes in various reliability and operational indicators on the values of indicators describing the states of the analysed railway video surveillance system. System repair and damage intensities shown in Table 1 were adopted for the analysis. The adopted values were calculated based on [19–22], as well as operational data provided by the Department of Operation and Passenger Service at PKP Polskie Linie Kolejowe S.A.

Table 1. System reliability parameters.

Parameter	Value (1/h)
λ_{12}	0.00001
λ_{13}	0.00002
λ_{14}	0.000025
λ_{15}	0.000004167
λ_{23}	0.00002
λ_{34}	0.000025
λ_{45}	0.000004167
μ_{21}	0.0208
μ_{31}	0.0416
μ_{41}	0.125
μ_{51}	0.5

By adopting Equations (13)–(17), and by applying the reverse Laplace transform and the values from Table 1, we obtain the following probabilities for the tested system to stay in individual operating states, for an exponential distribution:

- duration of the railway video surveillance system test—1 year:

$$t = 8760 \ (h) \tag{18}$$

- probability of the tested video surveillance system remaining in the state of full fitness S_1 for a period of 1 year:

$$R_0(t) = 0.9988319222061706 \tag{19}$$

- probability of the tested video surveillance system remaining in the state of partial fitness S_2 for a period of 1 year:

$$Q_2(t) = 0.00047974636032957274 \tag{20}$$

- probability of the tested video surveillance system remaining in the state of partial fitness S_3 for a period of 1 year:

$$Q_3(t) = 0.00048014975066258263 \tag{21}$$

- probability of the tested video surveillance system remaining in the state of partial unfitness S_4 for a period of 1 year:

$$Q_4(t) = 0.00019985575200001795 \tag{22}$$

- probability of the tested video surveillance system remaining in the state of full unfitness S_5 for a period of 1 year:

$$Q_5(t) = 8.32593 \cdot 10^{-6} = 0.000008325930837503415 \tag{23}$$

A graph of changes in the probability of the analysed railway video surveillance system remaining in the state of full fitness S_1 for a period of 1 year is shown in Figure 6.

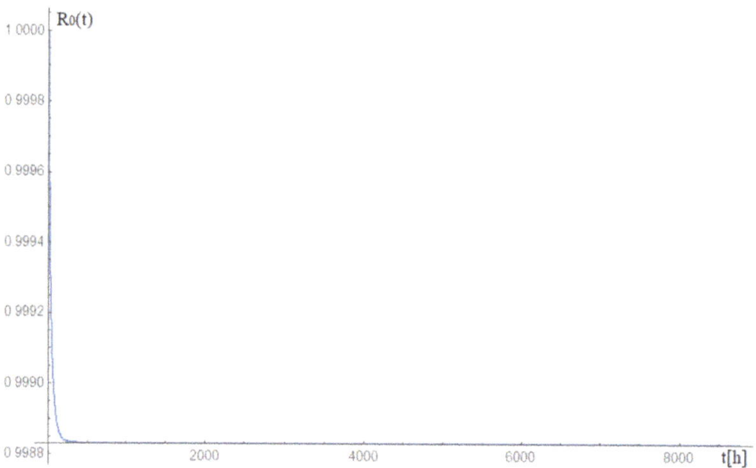

Figure 6. Graph of changes in the probability of the analysed railway video surveillance system remaining in the state of full fitness S_1 for a period of 1 year. Source: own study.

Assuming that the time of restoring the analysed system to the state of full fitness $\mu_{51} = t_{51} - 1$ (h) is confined within a limited range, $t_{51} \in \langle 12; 168 \rangle [\,h]$. This means, that within 0.5 to 7 days, it is probable that the analysed CCTV system will find itself in the state of full fitness, which is shown in Figure 7.

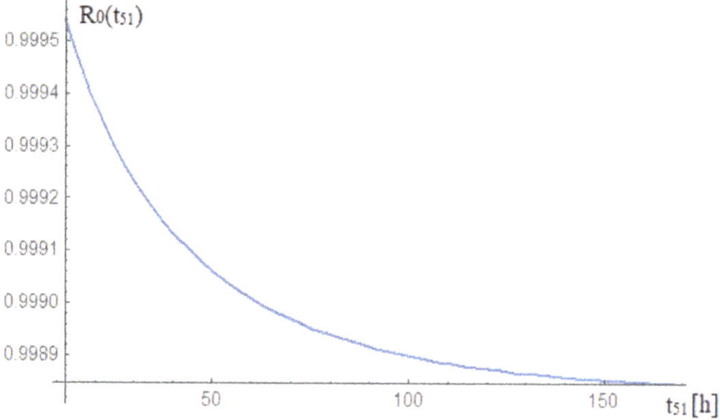

Figure 7. Dependence between the probability of the analysed railway CCTV system staying in the state of full fitness during the restoration of full system fitness. Source: own study.

4. Conclusions

The centralized video surveillance (CCTV) system described in the article is one of the largest solutions of this type under design within the European Union. Current solutions based on an integration platform usually involve single lines or a group of lines with limited territorial coverage (e.g., urban underground).

A centralized CCTV (video surveillance) system satisfies all the requirements that the national railway infrastructure administrator adopted in its regulations and documents that constitute the system basis, i.e., [19–23]. Based on the graph in Figure 7, it is possible to rationalize the actions associated with the possible restoration of full fitness of the analysed railway CCTV system. The time to restore the analysed system to a fully operational state is within a limited range. The analysed system is in a fully operational state for up to 7 days. The conducted analysis covering the railway CCTV system enables one to assess the security level of the solutions applied therein. A similar operational analysis, but based on the example of GPS receivers, has been conducted and described in [24]. The obtained result, i.e., a ~99.88% probability for a system to stay in the state of full fitness throughout a year, means that the centralized video surveillance system exhibits availability similar to that of other complementary railway systems [25–27], e.g., the GSM-R network (99.95), with 10.5 h downtime per year. Due to the scale and complexity of the system, this is a good result.

This enables one to improve the values of reliability indicators and to rationalize the operation of the tested system. According to [23], further improvement of reliability indicators should be based on increasing the reliability of systems that are not currently redundant (e.g., access switches at railway facilities), and shortening the response times (failure notification and repair) required by [19–22], which will translate to reducing the MTTR Mean Time To Repair) parameter, hence improving system reliability. Increasing the requirements in terms of repair speed must be preceded by a profitability analysis; it may entail significantly higher costs.

The resistance of the constructed system to factors not associated with operational damage to the elements will be a separate issue. Both the GS and CBIP are located in the same place (Poznań), and their functions cannot be taken over by any other unit. Therefore, there is a risk that in the event of a natural disaster or a terrorist attack, the centralized CCTV system will remain completely unfit for a prolonged period. It may be justified to implement a redundant server room and service centre located elsewhere, and not in Poznań. Other than that, operational damage was not taken into account in the calculations conducted for the purposes of this article.

The aforementioned article completely omits issues associated with the MPLS backbone network reliability. The analysis of this network's reliability will be the subject matter of another paper. The omission was due to the high reliability resulting from the network structure, i.e., redundant multiple-ring connections. According to the information received from the Department of Operation and Passenger Service at PKP Polskie Linie Kolejowe S.A., the IP MPLS is characterized by reliability exceeding the CCTV system.

Author Contributions: Conceptualization, P.S.; methodology, Z.K., P.S., M.S. and M.R.; validation, Z.K., P.S., M.S. and M.R.; formal analysis, Z.K., P.S., M.S. and M.R.; investigation, Z.K., P.S., M.S. and M.R.; resources, Z.K., P.S., M.S. and M.R.; data curation, P.S. and Z.K. writing—original draft preparation, P.S. and Z.K.; writing—review and editing, P.S. and Z.K.; visualization, P.S. and Z.K.; supervision, P.S. and Z.K.; project administration, P.S. and Z.K. All authors have read and agreed to the published version of the manuscript.

Funding: This research received no external funding.

Institutional Review Board Statement: Not applicable.

Informed Consent Statement: Not applicable.

Data Availability Statement: Not applicable.

Conflicts of Interest: The authors declare no conflict of interest.

References

1. Garlikowska, M.; Gondek, P. Sytuacja podróżnych na polskich dworcach kolejowych w aspekcie bezpieczeństwa z uwzględnieniem rozwiązań dla osób niepełnosprawnych. *Prace Instytutu Kolejnictwa* **2017**, *153*, 23–28.
2. Priks, M. The Effects of Surveillance Cameras on Crime: Evidence from the Stockholm Subway. *Econ. J.* **2015**, *125*, 289–305. [CrossRef]
3. Zarębski, A.; Szmigiel, P. *Wytyczne Dotyczące Projektowania i Budowy Systemów Monitoringu Wizyjnego (SMW) na Obiektach Obsługi Pasażerskiej Ipi-4*; PKP Polskie Linie Kolejowe S.A.: Warszawa, Poland, 2020.
4. Kim, J.; Park, N.; Kim, G.; Jin, S. CCTV video processing metadata security scheme using character order preserving-transformation in the emerging multimedia. *Electronics* **2019**, *8*, 412. [CrossRef]

5. Brahim, M.B.; Mir, Z.H.; Znaidi, W.; Filali, F.; Hamdi, N. QoS-Aware Video Transmission Over Hybrid Wireless Network for Connected Vehicles. *IEEE Access* **2017**, *5*, 8313–8323. [CrossRef]

6. Mina Qaisar, S.; Sidiya, D.; Akbar, M.; Subasi, A. An Event-Driven Multiple Objects Surveillance System. *Int. J. Electr. Comput. Eng. Syst.* **2018**, *9*, 35–44. [CrossRef]

7. Chen, S.; Yu, S.; Lü, J.; Chen, G.; He, J. Design and FPGA-Based Realization of a Chaotic Secure Video Communication System. *IEEE Trans. Circuits Syst. Video Technol.* **2018**, *28*, 2359–2371. [CrossRef]

8. Chumuang, N.; Ketcham, M.; Yingthawornsuk, T. CCTV based surveillance system for railway station security. In Proceedings of the International Conference on Digital Arts, Media and Technology (ICDAMT), Phayao, Thailand, 25–28 February 2018; pp. 7–12. [CrossRef]

9. Hadiwardoyo, S.A.; Gao, L. Integrating a Middleware DDS Application for Safety Purposes in an Underground Railway Environment. In Proceedings of the 3rd International Conference on Computer and Communication Systems (ICCCS), Nagoya, Japan, 27–30 April 2018; pp. 46–50. [CrossRef]

10. Punn, N.S.; Agarwal, S. Crowd Analysis for Congestion Control Early Warning System on Foot Over Bridge. In Proceedings of the 2019 Twelfth International Conference on Contemporary Computing (IC3), Noida, India, 8–10 August 2019; pp. 1–6. [CrossRef]

11. Park, H.; Park, S.; Joo, Y. Robust Detection of Abandoned Object for Smart Video Surveillance in Illumination Changes. *Sensors* **2019**, *19*, 5114. [CrossRef] [PubMed]

12. Wang, Y.; Zhu, L.; Yu, Z.; Guo, B. An Adaptive Track Segmentation Algorithm for a Railway Intrusion Detection System. *Sensors* **2019**, *19*, 2594. [CrossRef] [PubMed]

13. Ullah, F.U.M.; Ullah, A.; Muhammad, K.; Haq, I.U.; Baik, S.W. Violence Detection Using Spatiotemporal Features with 3D Convolutional Neural Network. *Sensors* **2019**, *19*, 2472. [CrossRef] [PubMed]

14. Rodolfo, Q.; Darwin, T.; Adín, R.; Helio, P. Multi-Stream Networks and Ground Truth Generation for Crowd Counting. *Int. J. Electr. Comput. Eng. Syst.* **2020**, *11*, 33–41. [CrossRef]

15. Gajbhiye, P.; Naveen, C.; Satpute, V.R. VIRTUe: Video surveillance for rail-road traffic safety at unmanned level crossings. In Proceedings of the 2017 IEEE Region 10 Symposium (TENSYMP), Cochin, India, 14–16 July 2017. [CrossRef]

16. Paś, J.; Rosiński, A.; Białek, K. A reliability-operational analysis of a track-side CCTV cabinet taking into account interference. *Bull. Pol. Acad. Sci. Tech. Sci.* **2021**, *69*, e136747. [CrossRef]

17. Khac-Hoai Nam, B.; Hongsuk, Y.; Jiho, C. A Multi-Class Multi-Movement Vehicle Counting Framework for Traffic Analysis in Complex Areas Using CCTV Systems. *Energies* **2020**, *13*, 2036. [CrossRef]

18. Veena, B.M. Reliability Analysis in Telecommunications. *Not. Am. Math. Soc. Commun.* **2020**, *67*, 6. [CrossRef]

19. Alessio, G.; Sergio, S. AdViSED: Advanced Video SmokE Detection for Real-Time Measurements in Antifire Indoor and Outdoor Systems. *Energies* **2020**, *13*, 2098. [CrossRef]

20. Zarębski, A.; Szmigiel, P. *Wytyczne Dotyczące Projektowania i Budowy Systemów Monitoringu Wizyjnego (SMW) na Obiektach Obsługi Pasażerskiej Ipi-5*; PKP Polskie Linie Kolejowe S.A.: Warsaw, Poland, 2020.

21. Zarębski, A.; Szmigiel, P. *Appendix No. 4 (Platform integrating PSIM (Physical Security Information Management)) to OPZ for Unlimited Tender on: Zaprojektowanie, dostawa i instalacja elementów prezentacji dynamicznej informacji pasażerskiej oraz systemu monitoringu wizyjnego wraz z infrastrukturą techniczną na dworcach, stacjach i przystankach kolejowych w ramach realizacji projektu pn. Projekt, dostawa i instalacja elementów prezentacji dynamicznej informacji pasażerskiej oraz systemu monitoringu wizyjnego wraz z infrastrukturą techniczną na dworcach, stacjach i przystankach kolejowych*; PKP Polskie Linie Kolejowe S.A.: Warsaw, Poland, 2018.

22. Zarębski, A.; Szmigiel, P. *Appendix No. 7 (Guidelines for a telecommunications container for the needs of GS1 server rooms) to the OPZ for Unlimited Tender for: Zaprojektowanie, dostawa i instalacja elementów prezentacji dynamicznej informacji pasażerskiej oraz systemu monitoringu wizyjnego wraz z infrastrukturą techniczną na dworcach, stacjach i przystankach kolejowych w ramach realizacji projektu pn. Projekt, dostawa i instalacja elementów prezentacji dynamicznej informacji pasażerskiej oraz systemu monitoringu wizyjnego wraz z infrastrukturą techniczną na dworcach, stacjach i przystankach kolejowych*; PKP Polskie Linie Kolejowe S.A.: Warsaw, Poland, 2018.

23. Zarębski, A.; Szmigiel, P. *Wymagania na transmisję danych systemów SMW, SPA i SDIP oraz integrację z siecią teletransmisyjną PKP Polskie Linie Kolejowa S.A. Ie-122*; PKP Polskie Linie Kolejowe S.A.: Warsaw, Poland, 2019.

24. *PN-EN 62676-1-1 Systemy dozorowe CCTV stosowane w zabezpieczeniach—Część 1-1: Wymagania systemowe*; PKN Poland: Warsaw, Poland, 2014.

25. Rychlicki, M.; Kasprzyk, Z.; Rosiński, A. Analysis of Accuracy and Reliability of Different Types of GPS Receivers. *Sensors* **2020**, *20*, 6498. [CrossRef] [PubMed]

26. Malfait, W.; Domingo, B.; Spaans, C.; Holvad, T. Operational Requirements Of Railway Radio Communication Systems. Available online: https://www.era.europa.eu/sites/default/files/library/docs/ex_post_evaluation/era_ee_ex_post_operational_radio_requirements_en.pdf (accessed on 20 October 2021).

27. Szmigiel, P.; Kasprzyk, Z. Assessment of Availability of Railway Video Surveillance System. In *Research Methods and Solutions to Current Transport Problems*; Springer: Cham, Switzerland, 2020; pp. 430–439. [CrossRef]

Article

Assessment of the Reliability of Wind Farm Devices in the Operation Process

Stanisław Duer [1,*], Jacek Paś [2], Aneta Hapka [3], Radosław Duer [3], Arkadiusz Ostrowski [4] and Marek Woźniak [4]

1   Department of Energy, Faculty of Mechanical Engineering, Technical University of Koszalin, 15-17 Raclawicka St., 75-620 Koszalin, Poland
2   Faculty of Electronic, Military University of Technology of Warsaw, 2 Urbanowicza St., 00-908 Warsaw, Poland; jacek.pas@wat.edu.pl
3   Faculty of Electronic and Informatics, Technical University of Koszalin, 2 Sniadeckich St., 75-620 Koszalin, Poland; aneta.hapka@tu.koszalin.pl (A.H.); radoslaw.duer@wp.pl (R.D.)
4   Doctoral School, Technical University of Koszalin, 2 Sniadeckich St., 75-620 Koszalin, Poland; arkadiusz.ostrowski@s.tu.koszalin.pl (A.O.); marek.wozniak@s.tu.koszalin.pl (M.W.)
*   Correspondence: stanislaw.duer@tu.koszalin.pl; Tel.: +48-943-478-262

Abstract: The article deals with simulation tests on the reliability of the equipment of the wind farm WF in the operation process. The improvement, modernization, and introduction of new solutions that change the reliability, as well as the quality and conditions of use and operation of wind farm equipment, require testing. Based on these tests, it is possible to continuously evaluate the reliability of the equipment of WF. The issue of reliability assessment of wind farm equipment, for which intelligent systems, diagnostic systems DIAG, and Wind Power Plant Expert System (WPPES) are used to modernize the operation process, can only be tested in a simulative way. The topic of testing the reliability of complex technical objects is constantly developing in the literature. In this paper, it is assumed that the operation of wind farm equipment is described and modeled based on Markov processes. The adoption of this assumption justified the use of the Kolmogorov–Chapman equations to describe the developed model. Based on this equation, an analytically developed model of the wind farm operation process was described. The simulation analysis determines the reliability of the wind farm in terms of the availability factor $K_g(t)$. The simulation tests are performed in two phases using the computer program LabView. In the first stage, the reliability value in the form of the readiness factor $K_g(t)$ as a function of changes in the mean repair time value ranging {from 0.3 to 1.0} was investigated. In the second stage, the reliability value of WF devices was examined as a function of changes in the value of the average time between successive failures, ranging from 1000 to 3000 (h)}.

Keywords: reliability; servicing process; intelligent systems; wind farm device; diagnostic process; expert system; knowledge base; neural networks; diagnostic information

Citation: Duer, S.; Paś, J.; Hapka, A.; Duer, R.; Ostrowski, A.; Woźniak, M. Assessment of the Reliability of Wind Farm Devices in the Operation Process. *Energies* **2022**, *15*, 3860. https://doi.org/10.3390/en15113860

Academic Editors: Paweł Ligęza and Silvio Simani

Received: 30 March 2022
Accepted: 28 April 2022
Published: 24 May 2022

Publisher's Note: MDPI stays neutral with regard to jurisdictional claims in published maps and institutional affiliations.

1. Introduction

Wind farm power equipment (wind power plant, unit transformers, etc.) in wind farms are technical facilities subject to the process of their continuous use or that remain ready for operation. Maintaining the technical condition of these devices to a good level of their operational properties requires an application of a specific operational policy (strategy). The idea of this wind farm equipment renewal strategy consists in minimizing the costs of the repair process or restoration of operational features and reducing the duration of the regeneration process of these devices. Developing an appropriate policy for the operation of wind farm equipment is a costly and difficult task. The authors' team has created a SERV expert program with the goal of assisting in the organization of the process of regeneration of operational features of wind farm equipment in the defined maintenance system. The SERV system is a large expert computer program that builds the structure of a system that regenerates wind farm equipment based on diagnostic data from these devices. The

95

SERV program determines the functional pieces of the wind farm equipment that need to be regenerated concerning their technical state in the first step. The SERV program then allocates specified sets (collections) of technical (regenerating) tasks to selected pieces of wind farm equipment at the following level. The SERV system optimizes the costs and the regeneration time of the wind farm equipment in the maintenance system. Understanding the quality of the operation process of the WF Wind Farm equipment, and thus learning about the reliability of the WF wind farm equipment, is the main research objective set out in this article.

The article presents the issues of simulation testing of the reliability of the WF Wind Farm equipment in its operation process. The problem of studying the operation process of complex technical facilities, including wind farm equipment with wind power plant WPP, and electrical subsystems (the power substation) is an important cognitive issue. This problem is of particular importance to the owners and users of WFs. They should explain how to handle organizational and technical operations in the WF device technical maintenance system. Only a well-organized WF renewal system will allow these facilities to be used to their full potential. The creation of dependable and suitable procedures and policies for the functioning of WF devices is the consequence of this sort of research work. The aforementioned difficulties have not been addressed in such a thorough manner in the literature.

The article covers the problem of simulation testing of the quality of the operation process: regeneration of the operational properties that improve the reliability of WF Wind Farm equipment. The issues presented in the article will be solved as follows. The second part of the article will present the methodology used for testing the reliability of the WF Wind Farm equipment based on the quality assessment of the regeneration process. The third part of the article will cover the issues related to the understanding and description of the operation (i.e., use and maintenance) of Wind Farm equipment. To organize simulation studies, the article presents and describes a model of the operation process of Wind Farm equipment where the intelligent SERV program was used. Another issue discussed in this part of the article will be related to an assessment of the reliability of WF devices after the use of intelligent regeneration systems. Presentation of those issues that improve the reliability of WF Wind Farm equipment constitutes an essential part of this section of the article. This problem is the main research goal of the article. The fourth section of the article is the main research part. The research conducted, which is covered in the article, concerns two issues:

Testing and evaluation of the reliability of Wind Farm devices in the operation process in relation to a reduction of the value of the repair time.

Testing and evaluation of the reliability of Wind Farm devices in the operation process in relation to changes in the value of the time required between the successive failures of Wind Farm devices.

The fourth section will cover the results and their analysis in the aspect of testing the quality of the operation process: regeneration of the operational features that increase the reliability of Wind Farm WF equipment.

Complex technical objects used in the exploitation process lose their functional properties. Their ability to perform the required function (their tasks as intended) is diminished. In literature [1–9], the problem related to determining the utility is called a functional resource. The decrease in the operational capacity is also directly related to the decrease in the reliability of technical facilities, including the wind farm equipment WF. The decrease in the reliability of technical objects results mainly from aging changes and the unfavorable influence of external factors. Therefore, there arises the problem of measuring and determining the current reliability of Wind Farm WF devices and restoration of functional properties of complex technical devices. The problem is quite complicated for technical devices that perform their tasks continuously, such as wind farm devices, medical devices, and others.

The designed system of an automatic regeneration of the functional properties of objects forms the bases for optimization of costs connected with prevention activities. This system fully minimizes the costs connected with the organization of the maintenance system of an object. The regeneration of the object takes place at the time when it is required. This is ensured by an intelligent diagnostic system of the object which is constructed based on an artificial neural network, especially such a network which reliably and credibly recognizes the states of the object for which prevention activities need to be performed [10–20]. There are no losses: no costs connected with ineffective use of the object, which may occur during operation when the object is not fit or it is in the state of incomplete fitness. This system eliminates the costs connected with the regeneration of those elements of the object which do not require it and are in the state of fitness. The designed intelligent maintenance system (including the intelligent diagnostic system) of the object ensures the regeneration of those internal (constructional) elements which require this, are in the state of incomplete fitness {1} or unfitness {0}.

The study by Kacalak et al., and others [21–24], provides an overview of an effective measuring track, which is a key component of a diagnostic system's structure. Furthermore, theoretical foundations for developing a measuring system using a computer measuring card were presented, with the goal of creating a measuring database for the diagnostic system. The study's findings were backed up by an example of information measuring database for the item in question. The studies address challenges connected to the automation of technical processes and the application of human knowledge in the development of intelligent systems for the purposes of diagnostic testing of technical items [25–27].

Technical diagnostics of technical devices is another essential problem that forms the basis of the organization of technical operations. The diagnostic examinations of devices are oriented towards the examination and identification of the technical state of the object examined. In the diagnostics of technical devices, the recognition of states in bivalent and trivalent logic is used. In the organization of the operation process, which renews the technical object, diagnoses determined by the diagnostician using trivalent logic are of the greatest practical significance. The studies by Zurada and Duer [28,29] constitute the canon of achievements in this area.

In the paper [30], the authors presented, inter alia, the essence, and methodology of developing models of the operation process of complex technical objects. In this work, the authors present the problem of a qualitative assessment of a maintenance process organized in this manner is the objective of this article. For this purpose, a program of simulation investigations was presented in the article. The research program consists of a description of the models of the operation processes of technical objects, determination of the input data to the investigations, which are the quantities of the operation time of a technical object being the summary duration time of the regeneration (repairs) and the use of objects and the determination of the indexes of a qualitative assessment of the regeneration of an object in the operation process. The results of the study were justified with an example of simulation investigations concerning the effects of the operation process with the regeneration of a technical object in an intelligent system with an artificial neural network.

The study by Dyduch and Siergiejczyk et al. includes a description of reliability–exploitation analysis is vital [31–33]. Electromagnetic compatibility of applied electrical and electronic devices [34] is equally relevant, but this aspect is not scrutinized in this article. Yet one must not ignore the influence of electromagnetic interference on the functioning of electronic devices [35–45].

As in reliability research it is important to model the technical object itself and its operation process. There is an important research issue in determining the reliability of wind farm equipment. The issues of graphical and analytical modeling to assess the reliability of technical objects are presented in the works of Siergiejczyk and others [34–36,38–40]. Models of the operation process of technical objects based on the theory of Markov processes are particularly important in the theory and practice of reliability of technical objects.

The Kolmogorov–Chapman equation is used in the non-reliability assessment of technical objects in these models. This type of research approach is also presented in this article.

Another direction of reliability testing in the operation process of technical objects and systems is the use of Chapman–Kolmogorov equations in them. This is particularly evident in the works by Siergieczyk and others [35,36,38–40]. This article describes the issues related to the analysis of reliability–exploitation of power supply systems in transport telematics systems PSSs in TTDs. This paper characterizes solutions, which are applied in supply systems and describes a PSS in a TTD from the main source and a standby one. This enables determining the dependencies denoting the probabilities of the system staying in full ability state, safety threat state, and safety unreliability state. Quality analysis of the PSS in TTD was conducted, and the indicator value of the supply continuity quality was evaluated. This indicator allows the demonstration of continuity quality of power supply CQoPS dependency on many quality dimensions, not just reliability. An example demonstrates the calculation of the CQoPS factor for both the main and the standby power supply employing three observations each influencing the quality. The presented considerations in the field of quality and reliability–exploitation modeling of PSS can be applied as well in other public utility facilities (including those classified as critical infrastructure). The character of functions performed by critical infrastructure.

The issues of the modeling of the operation process of technical objects are presented in the following publications by Nakagawa and others [41–44]. The author's research is also significant. In these articles, a mathematical approach to simulating this process is offered. In the operation process, the author interprets the states of the object and the essence of changes (transitions) between them. A presentation (use) of an approach to the organization of the object's renewal process in the maintenance system is an important element in the modeling of the object's operation process. The use of the current state of the object in the structure of the operation process, also known as the operation of the object according to its state, is a new approach created in the author's investigations.

Issues regarding the use and operation of electrical equipment in wind farms and wind farms are presented by Badrzadeh et al., Pogaku et al., and others [5,7,42–45]. The works on the construction, functioning, and modeling of electrical devices located in wind farms are well developed in these works.

An essential element in the modeling of the operation process for the complex technical object is the development of a model of the renewal process for an intelligent maintenance system. These issues are presented in publications including those by Buchannan et al., Duer [11,13]. In his studies, the author presents issues related to the determination of the systems maintenance models. For this purpose, the form (dimension) of the matrix of the object's structure is accepted. It is transformed into the form of the object's maintenance matrix. Elements in the maintenance matrix are assigned to the primary elements of the object. The elements of the object's maintenance matrix describe explicitly the subsets of those technical and technological activities which must be performed upon a given element of the object for its renewal. The process of assigning the elements of the object's structure to adequate renewing activities with the use of appropriate materials and resources is a complicated task. These issues are continuously being developed and improved in the author's studies.

The current research work supporting the development of expert and advisory systems is focused on the issues related to the improvement of methods of acquiring specialist knowledge of a person, with the wide participation of modern solutions such as artificial intelligence and intelligent systems. This issue is presented in the works [16,17]. The issues of testing the reliability of wind farm equipment in its operation process are presented graphically in Figure 1.

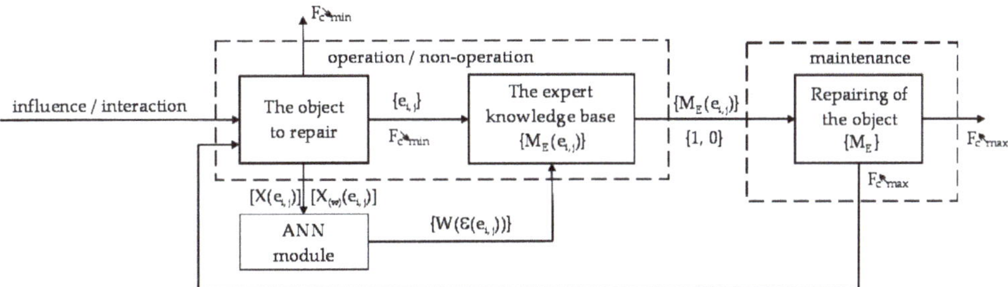

Figure 1. Diagram of operation process for technical object utilizing the artificial neural network.

Where the following stand for:

-   $X(e_{i,j})$ is the diagnostic signal in the j-th element of the i-th set,
-   $X_{(w)}(e_{i,j})$ is a model signal for $X(e_{i,j})$ signal,
-   F_C is min. or max. value of the function of the use of the object,
-   $W(\varepsilon(e_{i,j}) = \{3, 2, 1, 0\})$ is the diagnostic information-value of state assessment logics for element "j" within "i" module of the object.

Getting to know the current level of reliability of the devices used by the Wind Farm WF and other complex technical facilities is possible through their diagnosis. Diagnosis using inference (state recognition) in multi-valued logic is particularly useful [25]. At present, there is widespread progress in the development of specialized diagnostic devices. The issues are presented in the works [26,27]. It is particularly visible in the diagnostics of medical devices, energy technology, etc. However, these are diagnostic devices with an individual application for the selected device under test. There is no diagnostic device on the market with a wide (general) spectrum of practical diagnostic use. The works by Duer and others show that diagnostic devices represent a common and uniform technical solution. It is a modular solution with the following functional elements: measurement, diagnostic, and diagnostic knowledge base. For each diagnosed device or technical or technological process, only the measurement knowledge base and the measurement system, acquisition, etc., are relevant.

It is essential to comprehend the structure and principles of operation of technological devices and inaccurately diagnose them. Concerns relating to the operation of wind farm equipment are discussed in the following works [16].

In the work by Duer [45], he presented research on the reliability of Wind Farm equipment based on analytical models using reliability dependencies. However, the obtained results indicate that this approach is quite difficult to implement in simulation studies. The article presents the organization, implementation, and analysis of the simulations carried out for the evaluation of the quality of the maintenance system of wind farm equipment WF. The important aspect for the reader is to present models of wind farm equipment WF operation processes. The reader will find the issues of building and organizing the operation process of complex technical objects in [4]. Three models of wind farm equipment operation processes WF were used for the simulations. The model is Model A, an operation process of a wind power plant that uses an intelligent maintenance system with an artificial neural network. The second model is Model B, an operation process of the object which uses information in bivalent logic, a model with a maintenance system organized by planning its optimal prevention activities. The third is Model C, an operation process of a wind power plant with a maintenance system that is organized classically without any examination of the state in the assessment process: a strategy for object's maintenance is based on manual planning of prevention activities and arbitrary operator's selection of its scope.

The issues related to the description and testing of the individual elements that describe the operation process of technical facilities are well presented in publications. However, there are no studies that present the challenges of research into and organization of the operation process of complex technical facilities in a comprehensive manner. As a result, the article aims to model the reliability of WF Wind Farm equipment during its operation. This task will require solving the research problems presented below. The first issue is related to understanding and describing the problems of the diagnostics of WF devices. Another problem is understanding and describing the operation process (i.e., the use and maintenance) of Wind Farm devices. An important issue presented in the article is understanding and description of the organization of the technical maintenance system in the process of the operation of the facility tested.

The research on the reliability of wind farm technical devices with the use of intelligent systems was presented methodologically. For the WF operation process described in this way, its model was developed, and analytical relationships were determined on the basis of which the tested reliability values are determined in the form of the availability function ($K_g(t)$).

The article deals with the problem of testing the reliability of WF depending on the influence of one of the parameters on this value, i.e., the average time between successive failures. The application of this research parameter to the reliability of WF in such a way has not yet been undertaken in publications. The novelty of this article is also its use as a research tool in the form of the LabView computer program. The obtained results from the simulation tests are interesting and were not presented in the publications in the form presented in this article. The task of understanding the reliability of the Wind Farm equipment has become the main research goal presented in this article.

2. Methodology for Testing the Reliability of the Wind Farm Equipment in the Operation Process

Each study of any technical object, and even more so a simulation study, requires input data characterizing the actual operation process of the selected object class and its simulation models (Figure 2). The study of the actual operation of the facility is the basis for obtaining data for the simulation study of process models.

The required input data for the tests are the following quantities:

- the time of use of the object T is the time the object is in a fit condition,
- the time of removing the T_a object inoperability,
- time of performing preventive repair of T_p,
- period of projected (optimal) prevention θ^*,
- period of planned prophylaxis θ.

The source of the above data may be the observation of actual operation processes and a properly prepared and implemented simulation experiment. The results concerning the study of the actual operation process of various classes of technical objects are presented, among others, in the study. The simulation experiment consists of the following components (Figure 2):

- Model of the tested facility operation process,
- Test program,
- Research tools—the use of a computer in research,
- Analyze the obtained data.
- The study of the models of the object operation processes was carried out using the same test criteria test conditions, such as:
- Functions describing the object operation process and the inputs consumed,
- Input data characterizing the operation process of complex objects.

To determine the appropriate measure characterizing the quality of the facility operation process, other quantities describing the efficiency of wind farm equipment used in the operation process should be analyzed.

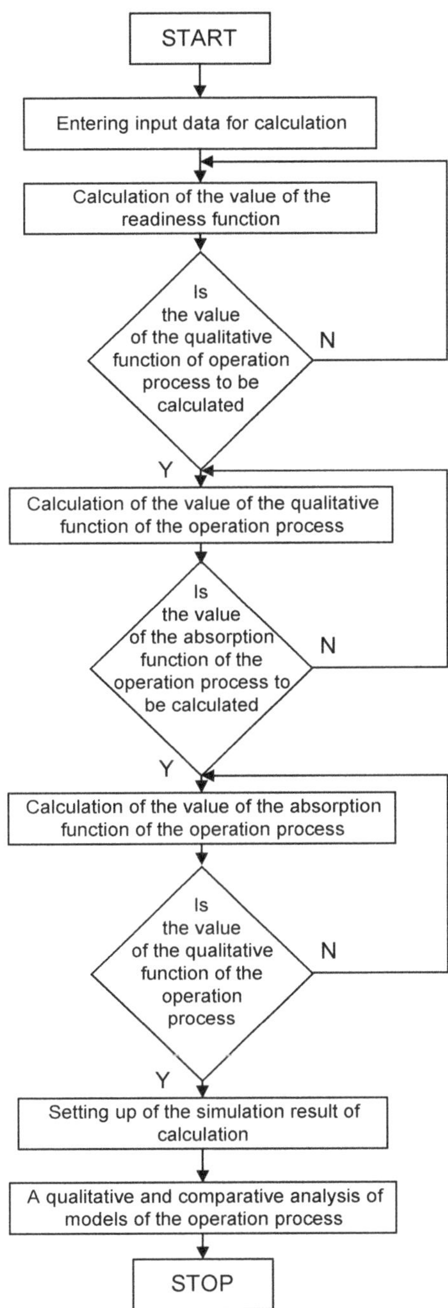

Figure 2. Algorithm of simulation investigations concerning the quality of the assessment of the operation process of the technical object.

The methodology of testing the reliability of the wind farm equipment operation process is an activity of research, analysis, and evaluation in this area. In order to better understand the research activities in the field of reliability assessment of the wind farm equipment operation process, a graphical diagram of this operation was developed (Figure 2). The essential points in this algorithm are as follows:

1. Understanding and describing the operation process (use and renovation) of wind farm equipment.
2. Development of models of the operation process of wind farm equipment.
3. Adoption of the quantity (function) characterizing the reliability test of the wind farm equipment operation process. For reliability tests, a reliability value known in the literature [2,38] was proposed, which is the function and the availability coefficient ($K_g(t)$):

$$K_g\ (t) = P[S(t)] \tag{1}$$

where the following stand for: $P[S(t)]$ is the probability that the technical object is in a fit condition.

1. Reliability simulation tests of the wind farm equipment operation process are carried out using the same computer program.
2. The same input data is used in the reliability simulation tests of the wind farm equipment operation process.
3. The results obtained from the simulation tests concerning the reliability of the operation process of wind farm equipment are presented graphically on common charts that present the quantities studied.

3. The Four-State Model of the Operation Process of Wind Farm Equipment

The analysis (Figure 1) shows that the wind farm equipment that is implemented (switched on) into operation is effectively used. When the wind farm devices are effectively used and the wind farm performs its functions as intended, then it is in the reliability state shown in Figure 2, which is called the effective SO usage state. With the time of use of a wind farm WF, its functional resource continues to decrease (reliability decreases). Measurement of the WF reliability level (value) is measured on an ongoing basis by diagnosing by the DIAG system. The DIAG system is an autonomous diagnostic device. The developed intelligent diagnostic system is a set of specialized technical devices along with a diagnostic card. The work of the DIAG program was verified on the basis of the diagnosis of functional devices of the Wind Farm. Problems of this type are quite modest in the literature [4].

Diagnostic information about the WF reliability level is assessed by the user on an ongoing basis. If the reliability of a wind farm drops below the acceptable level, the user (owner) of a wind farm decides on sending for renewal in an intelligent system. Therefore, such an event, when the WF cannot effectively carry out its tasks, is transferred to the reliability state in Figure 3, which is called the state of ineffective use–repair, restoration of utility properties S10. In Figure 1, the intelligent renewal (maintenance) system SERV is built by two subsystems: the subsystem of the expert maintenance knowledge base and the maintenance subsystem. The SERV system is an extensive expert system (computer program). The SERV system is responsible for determining the maintenance information on the basis of the diagnostic information obtained from the DIAG program. The developed SERV program is made on the basis of the presented stepwise algorithms for the process of renewing the functional properties of the wind farm. The developed solution of the SERV service system was subjected to practical action in the scope of determining the set of maintenance information for the renewal of the devices of the Wind Power Plant and other wind farms. Based on this information, the WF is renewed in the handling system. The renewed WF devices are returned to their use in the operation process (Figure 3).

During their operation, the WF devices require periodic and cyclical repair technical and technological works, such as inspections, and tests of wind power plant safety devices such as evacuation, crane, fire protection, and other devices. In a situation where the WF devices are subjected to repair and technological works, such a reliability state is called preventive shutdown (planned repair works) S1. During their effective use, WF devices are subject to constant supervision in the field of their safe use. One of the developed technical solutions that support the concept of a decision by the WF operator is the Wind Power Plant Expert system WPPEs expert program [4]. In situations threatening the safe operation of WF devices, the operator temporarily excludes WF from its use. Such a reliable state of WF presented in Figure 3 is called the state of ineffective use marked as S01.

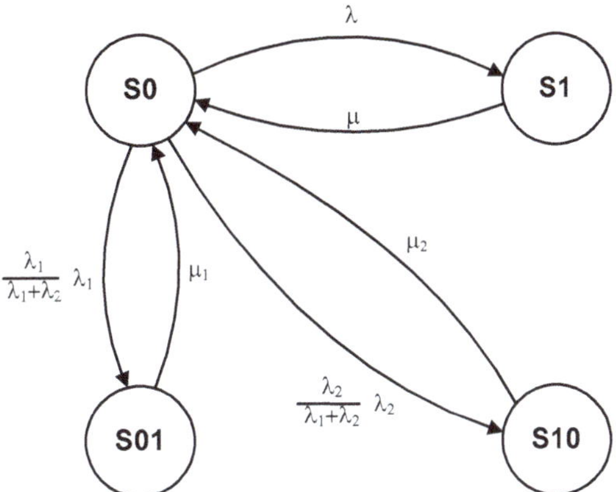

Figure 3. Diagram of the operation process of a Wind Equipment Farm not equipped with any intelligent expert systems SERV.

The problem related to the description of this type of exploitation process model was presented in the studies. The graphic form of this model is shown in Figure 3.

The analysis of the model of the operation process of the Wind Farm equipment presented in Figure 3 demonstrates that the facility may be in one of the following states:

- S0—effective use of the facility,
- S1—scheduled maintenance-preventive NP,
- S01—unscheduled maintenance,
- S10—ineffective use of the facility.

The transitions between the states in the model mean the following:

- λ—has an interpretation of the intensity of the system's transition from state S0 to state S1,
- μ—has an interpretation of the system's transition from state S1 to state S0,
- $\frac{\lambda_1}{\lambda_1+\lambda_2}\lambda_1$—has an interpretation of the intensity of the transition of the system from the state S0 to state S01,
- μ_1—has an interpretation of the intensity of the transition of the system from the state S01 to state S0,
- $\frac{\lambda_2}{\lambda_1+\lambda_2}\lambda_2$—has an interpretation of the intensity of the transition of the system from the state S0 to state S10,
- μ_2—has an interpretation of the intensity of the transition of the system from state S10 to state S0.

A technical facility used without the SERV expert system is damaged, and it modifies its place in the operating process. The object moves with intensity $\frac{\lambda_2}{\lambda_1+\lambda_2}\lambda_2$ from state S0 to the state of ineffective use S10. The time of the technical object remaining in the state of ineffective use S10 is determined by the random variable τ_{NA}. The value of the useful life of an ineffective technical facility results from the wind farm operating conditions or from changes in weather conditions. The technical object, once the unfitness has been located and removed that determines the cause of its ineffective use, is again transferred to the S0 use state with an intensity of μ_1. To identify the probability of the system remaining in the particular states, the network of crossings presented in Figure 4 shall be described with the following equations:

$$-\lambda \cdot P_0 + \mu \cdot P_1 - \lambda_1 \cdot \frac{\lambda_1}{\lambda_1 + \lambda_2} \cdot P_0 + \mu_1 \cdot P_{01} - \lambda_2 \cdot \frac{\lambda_2}{\lambda_1 + \lambda_2} \cdot P_0 + \mu_2 \cdot P_{10} = 0$$
$$\lambda \cdot P_0 - \mu \cdot P_1 = 0$$
$$\lambda_1 \cdot \frac{\lambda_1}{\lambda_1 + \lambda_2} \cdot P_0 + \mu_1 \cdot P_{01} = 0 \tag{2}$$
$$\lambda_2 \cdot \frac{\lambda_2}{\lambda_1 + \lambda_2} \cdot P_0 + \mu_2 \cdot P_{10} = 0$$

In the matrix notation, Relationship (2) can be presented as follows:

$$\begin{bmatrix} -\left(\lambda + \frac{\lambda_1}{\lambda_1 + \lambda_2} \cdot \lambda_1 + \frac{\lambda_2}{\lambda_1 + \lambda_2} \cdot \lambda_2\right) & \mu & \mu_1 & \mu_2 \\ \lambda & -\mu & 0 & 0 \\ \frac{\lambda_1}{\lambda_1 + \lambda_2} \cdot \lambda_1 & 0 & -\mu_1 & 0 \\ \frac{\lambda_2}{\lambda_1 + \lambda_2} \cdot \lambda_2 & 0 & 0 & -\mu_2 \end{bmatrix} \cdot \begin{bmatrix} P_0 \\ P_1 \\ P_{01} \\ P_{10} \end{bmatrix} = \begin{bmatrix} 0 \\ 0 \\ 0 \\ 0 \end{bmatrix} \tag{3}$$

By transforming Equation (3), the following relationships were obtained:

$$P_1 = \frac{\lambda}{\mu} \cdot P_0$$
$$P_{01} = \frac{\lambda_1}{\lambda_1 + \lambda_2} \cdot \frac{\lambda_1}{\mu_1} \cdot P_0 \tag{4}$$
$$P_{10} = \frac{\lambda_2}{\lambda_1 + \lambda_2} \cdot \frac{\lambda_2}{\mu_2} \cdot P_0$$

Obviously enough, it is known that the relationship is correct:

$$P_0 + P_1 + P_{01} + P_{10} = 1 \tag{5}$$

Therefore:

$$P_0 \cdot \left(1 + \frac{\lambda}{\mu} + \frac{\lambda_1}{\lambda_1 + \lambda_2} \cdot \frac{\lambda_1}{\mu_1} + \frac{\lambda_2}{\lambda_1 + \lambda_2} \cdot \frac{\lambda_2}{\mu_2}\right) = 1 \tag{6}$$

$$K_{g7} = K_0 = \frac{1}{\left(1 + \frac{\lambda}{\mu} + \frac{\lambda_1}{\lambda_1 + \lambda_2} \cdot \frac{\lambda_1}{\mu_1} + \frac{\lambda_2}{\lambda_1 + \lambda_2} \cdot \frac{\lambda_2}{\mu_2}\right)} \tag{7}$$

$$K_{g7} = K_0 = \frac{(\lambda_1 + \lambda_2) \cdot \mu \cdot \mu_1 \cdot \mu_2}{(\lambda_1 + \lambda_2) \cdot \mu \cdot \mu_1 \cdot \mu_2 + \lambda(\lambda_1 + \lambda_2) \cdot \mu_1 \cdot \mu_2 + \lambda_1^2 \cdot \mu \cdot \mu_2 + \lambda_2^2 \cdot \mu \cdot \mu_1} \tag{8}$$

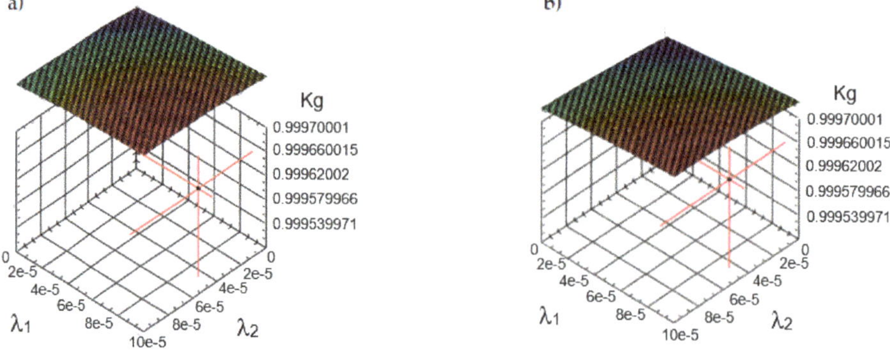

Figure 4. Graphs of the K_g availability coefficient in the process of testing the reliability of wind farm devices by changing the value of the average repair time for constant parameters, including: type I mean service time = 0.8 [h]; type II mean operating time = 1.3 [h]; mean time between successive failures = 1200 [h], where: (**a**) mean repair time 0.3 [h]; (**b**) average repair time 0.4 [h].

Using Expression (8), it is possible to determine the values that are of interest and related to the probabilities of a wind farm system remaining in its various operating states. If the assumption is made that the modeling of the operation process consists in determining the probabilities of a wind farm system remaining in individual states $\{S_0, S_1, S_{01}, S_{10}\}$, then the following values need to be determined:

- the likelihood function of the system being in a state S_0,
- the likelihood function of the system being in a state S_1,
- the likelihood function of the system being in a state S_{01},
- the likelihood function of the system being in a state S_{10}.

4. Results

4.1. Indices That Characterize the Process of Operating a Technical Facility

From the set illustrated in the literature [2,6,8,23,36] of these indicators characterizing the process of the operation of a technical object, the value best reflecting the operation process is the availability indicator K_g and the accessibility function $K_g(t)$. The process for calculating the availability feature $K_g(t)$ is generally simplified when calculated for a limit value at ($t \to \infty$). The size is closely related to the stationary characteristics of the damage and maintenance process. Due to this, the availability rate of K_g is the most appropriate measure to set out the efficiency of the operation process, which links both the utility and economic characteristics of the facility. The accessibility factor K_g of the object is the likelihood of the event that the object is operational after a sufficiently long period of operation ($t \to \infty$). The accessibility factor K_g determines the average proportion of the technical object's service life in the total service life, as represented by the following relationship:

$$K_g = \lim_{t \to \infty} K_g(t) = \lim_{t \to \infty} K_{gsr}(t) \tag{9}$$

where: $K_g(t)$ is the medium value of the accessibility factor K_g.

4.2. Testing and Evaluation of the Reliability of Wind Farm Devices in the Operation Process Due to the Decrease in the Value of the Repair Time

The organization of the wind farm equipment operation process is subject to continuous improvement and improving it in terms of its technical implementation as well as the methods and forms of its organization. One of the important aspects in the modernization and modernization of the renewal process of wind farm equipment is the use of the intelligent SERV system in it. The SERV system is a large computer expert program that generates a set of maintenance instructions based on the input diagnostic information about the tested object and the existing expert knowledge base in the SERV system's memory. The SERV system's established service knowledge base serves as the foundation for the development and implementation of a smart system for the renewal and implementation of the process of restoring functional qualities. The functional qualities of wind farm devices decline with use as a result of the implementation of tasks envisaged for them. The current level of ownership of the functional characteristics of wind farm devices is examined and determined during the diagnosis process. Hence, the diagnostic information determined during the diagnosis is presented in the form of a "WF devices status table". This information set, named the "Diagnostic Knowledge Base for WF Devices" DKB, is the input set for the SERV program. The SERV system develops a maintenance information set called the "Expert Knowledge Base for WF Equipment" EKB. The structure of EKB of WF devices is determined by the following sets of information:

- The service structure of the object is a set of information describing the internal structure of the object, which was determined on the basis of its functional model and the set of service information, adapted to the needs of the service. The service structure is determined by those basic elements of the facility that are in an incomplete or unfit condition and require renewal. The basic elements that require updating are called control elements. These elements are arranged in the facility's maintenance structure, which is defined by the i-th service levels, and each of the i-th levels by the j-th service layers;

- Methods of classifying (grouping) elements in the object's service structure, is a set of information describing the object's service structure due to the nature and tasks of the element, which was determined based on its functional model and the set of service information and others. The specialist assigns one s-th class from the set of classes to each operating element, where: s = {I to VIII}. Assigning a specific class to an element in a service structure is assigning an element to a specific group of devices, eg a class: mechanical, electrical, etc. The object's service structure adapted in this way is to adapt the nature of maintenance activities that are specific only to devices of a given group. This structure has a dimension like that of the operating structure, the elements describing this set of information are described through the i-th service levels, and each of the i-th levels is described by j-th service layers;

- Structures for renewing (servicing) maintenance elements is a set of maintenance information concerning the maintenance activities that must be performed in the process of renewing the maintenance elements of the facility. It is determined on the basis of assigning to each element of the maintenance structure a specific subset of maintenance activities, such as tuning, adjustment, maintenance, replacement, etc., selection of maintenance activities or their sets for individual operating elements are adapted to the current state, in which the structural elements of the facility are located; performing the required maintenance activities or their sets on individual maintenance elements will result in their transition to the state of suitability;

- Structure of service rules, this specialized set of object service information having the dimension of the internal structure of the object (levels × layers). This specialized maintenance knowledge base concerns the description of dependencies, rules, and relations in the scope of determining the information set of the object maintenance task, including structures: comparing the states of the object's elements, object maintenance, object maintenance activities, classifying object elements. The handling rules provide the user of the object with answers, on how and how to effectively organize the process of handling the object. The set of facility maintenance rules is the basis, apart from the algorithm, for designing an appropriate computer program that supports the effective restoration of the facility's operating features [4,45].

The renewal process of WF devices is organized on the basis of the set of maintenance information developed by the SERV system. The use of the SERV system in the process of refurbishing WF devices makes this type of refurbishment of this facility a modern solution, and at the same time innovative and effective. The issue of a refurbishment of wind farm devices is subject to continuous development to ensure that the restored usable resources of service elements have the character of a complete renovation, i.e., the same as the elements in the structure of a new facility (just implemented for use).

The problem of testing and assessing the reliability of wind farm devices in the operation process is a difficult task due to the decreasing value of the repair time. There is no description of this type of research in the literature. The article presents the thesis "Renewal of WF devices on the basis of information developed in the SERV system in a general manner and directly affects the reduction of the time of restoring the resource of functional properties (the object's presence in the maintenance system), and thus reducing the repair time" (Figure 1). The whole average duration of the renewal procedure (the object's stay in the handling system) is referred to as the average repair time. The article's research premise concerning the prospect of reducing the time it takes to repair wind farm devices stems from the essence of the facility renovation system, which is made possible by the usage of an intelligent SERV system. The intelligent SERV system used in the process of refurbishment of WF devices develops an effective "Knowledge base for servicing WF devices". In this set of maintenance information, the SERV system optimizes the service structure of WF equipment. The SERV system defines a set of operating elements of WF devices that have {3} states—the states of availability and do not require the need to restore the resource of operational properties. The size of the average repair time of WF devices is also influenced by the organization of the renewal process based on information from the SERV system.

On the basis of information from the SERV system, the service elements are renewed for the appropriate level based on their current state from the {3, 2, 1, 0} set. Then, the implementation of the renewal process (implementation of technical and technological activities) is also optimal. Only those technological activities (renewing service elements) are performed that are required (developed by the SERV system) to fully renew the elements of the WF devices. Therefore, the presented problems of organization and implementation of the wind farm equipment renewal process will result in reducing the repair time. The results obtained from the testing and evaluation of the reliability of wind farm devices in the operation process due to the decrease in the value of the repair time are presented in Table 1 and Figures 4–7.

The analysis of the test results presented in Table 1 and in Figures 4–7 shows that for the value of the mean repair time equal to (0.3), the reliability value determined in the study in the form of the readiness factor K_g (1) is 0.999793711044.

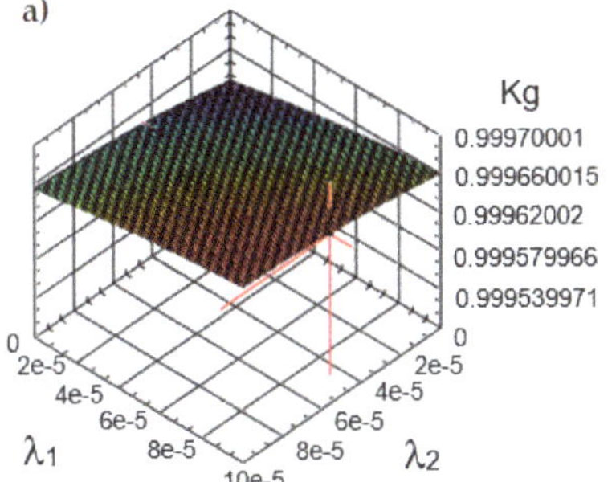
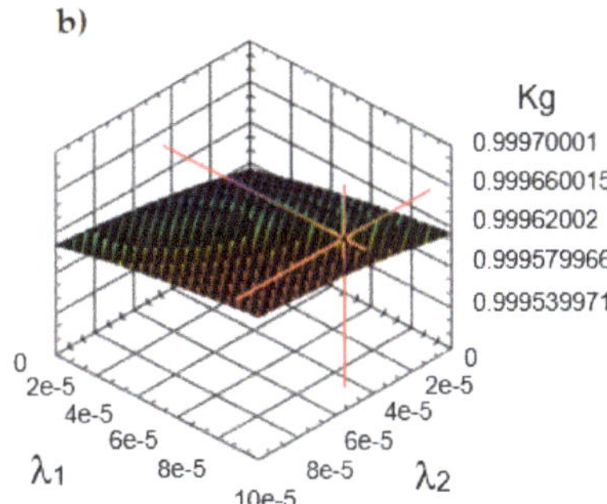

Figure 5. Graphs of the K_g availability coefficient in the process of testing the reliability of wind farm devices by changing the value of the average repair time for constant parameters, including: type I mean service time = 0.8 [h]; type II mean operating time = 1.3 [h]; mean time between successive failures = 1200 [h]. where: (**a**) average repair time 0.5 [h]; (**b**) average repair time 0.6 [h].

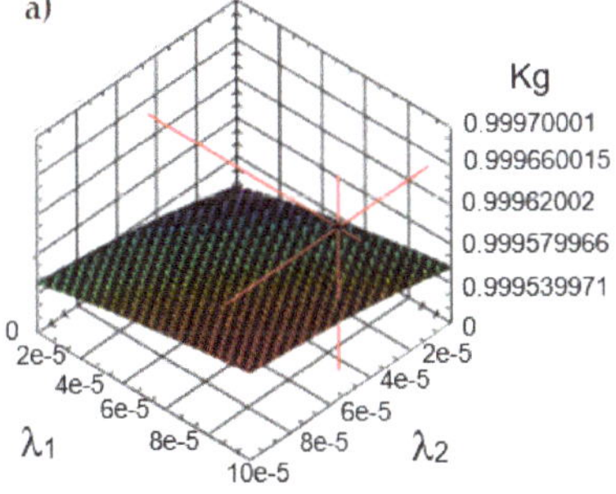
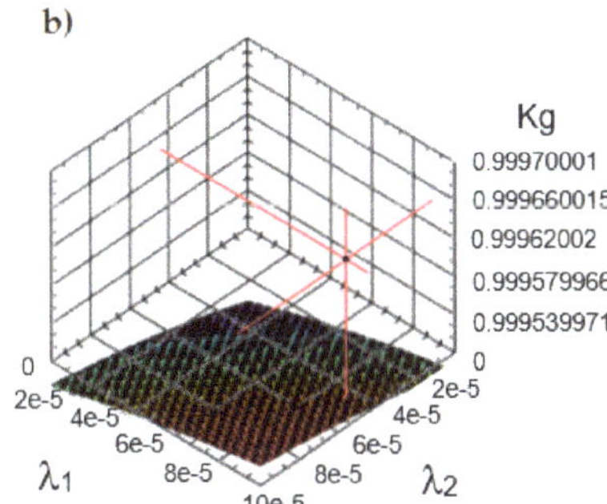

Figure 6. Graphs of the K_g availability coefficient in the process of testing the reliability of wind farm devices by changing the value of the average repair time for constant parameters, including: type I mean service time = 0.8 [h]; type II mean operating time = 1.3 [h]; mean time between successive failures = 1200 [h], where: (**a**) mean repair time 0.7 [h]; (**b**) average repair time 0.8 [h].

The assumed value of the average repair time of (0.3) means that in the maintenance system based on information from the SERV system, the renewal time of WF devices was reduced to 70% concerning the average value of the renewal time of wind farm devices implemented in the maintenance system organized traditionally (classic) [40,41]. The research verified (tested) the impact of the average repair time equal to the test value (1.0) appropriate for the renewal of wind farm devices in the maintenance system organized classically (without support from the use of intelligent systems). For the mean repair time value equal to (1.0), the determined reliability of wind farm devices renewed in the maintenance process in the form of the availability factor K_g (2) is 0.99943517 (Figure 8).

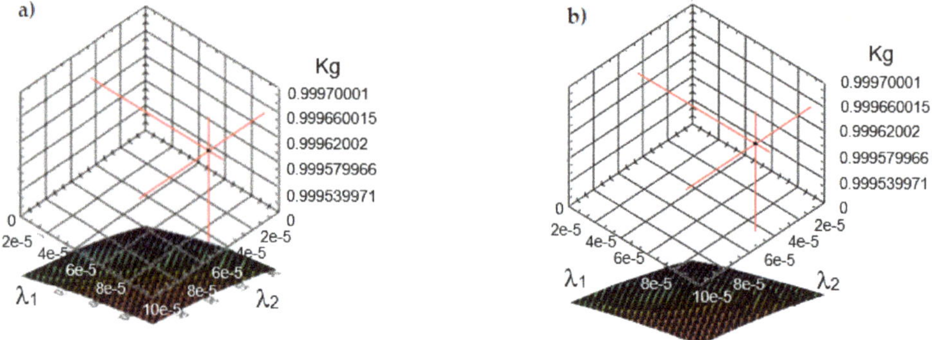

Figure 7. Graphs of the K_g availability coefficient in the process of testing the reliability of wind farm devices by changing the value of the average repair time for constant parameters, including: type I mean service time = 0.8 [h]; type II mean operating time = 1.3 [h]; mean time between successive failures = 1200 [h], where: (**a**) mean repair time 0.9 [h]; (**b**) average repair time 1.0 [h].

Table 1. Max. value readiness factor K_g.

Average Repair Time	Value Max. Readiness Factor K_g
0.3	0.999793711044
0.4	0.998742403291
0.5	0.987682914322
0.6	0.985590024178
0.7	0.974590023678
0.8	0.969544193147
0.9	0.969488234806
1.0	0.959435172789

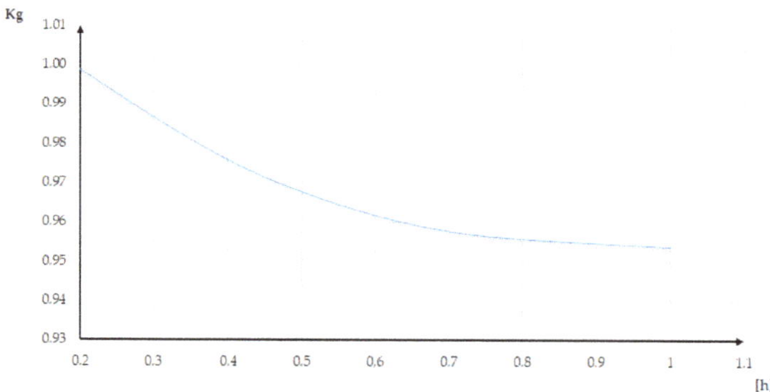

Figure 8. Graph of the reliability of wind farm devices in the operation process as a function of repair time.

The analysis of the K_g (1) and K_g (2) readiness factor values allows for determining the possible increase in the value in the form of the readiness factor ΔK_g on the basis of the relationship: $\Delta K_g = K_g$ (1)$-K_g$ (2), hence the value $\Delta K_g = 0.000358$. The analysis of the research results presented in (Figure 1) allows for the formulation of the following conclusions.

1. Renewal of wind farm devices in the maintenance system organized on the basis of information developed in intelligent systems, including: diagnostic (DIAG) and maintenance (SERV) significantly increases the reliability of the renovated facility.
2. The use of innovative intelligent systems in the renewal process of wind farm devices brings an increase in the reliability of the renewable wind farm devices in the form of the readiness factor value is $\Delta K_g = 0.000358$.
3. The intelligent SERV system is a good computer tool that effectively supports the construction and organization of maintenance systems renewing the wind farm equipment.

4.3. Testing and Evaluation of the Reliability of Wind Farm Devices in the Operation Process Due to Changes in the Value of Time between Successive Failures of Wind Farm Devices

It is important to test and evaluate the reliability of wind farm devices in the operation process due to changes in the value of time between successive failures of wind farm devices and a research aspect that is difficult to implement. The average value of the time between successive failures of wind farm devices in the operation process is one of the basic indicators in assessing its reliability in (Table 2) and (Figures 9–13). The average value of the time between successive failures of WF devices is directly related to the quality and accuracy of the device renewal, the method and strategy used in the renovation process, the full use of service information about the device, and other factors, and can be used to assess the reliability of WF devices refurbished in an intelligent maintenance system. As a result, the influence of the average value of time between successive failures on the reliability quality of wind farm equipment was evaluated in the next simulation study.

Table 2. Max. value readiness factor K_g.

Mean Time between Successive Failures (h)	Value of Max. Readiness Factor K_g
1000	0.999427552702
1200	0.989498551118
1300	0.979567543594
1500	0.969612736465
1800	0.959688899662
2000	0.959728217231
2500	0.949782318116
3000	0.949815173253

The study assumed the range of changes in the mean value of the time between successive lesions in the range of {1000 ÷ 3000} [h]. The results obtained from testing the average value of the time between successive failures of wind farm devices for the reliability of wind farm devices in the operation process are presented in Table 2 and Figures 9–12.

The analysis of the test results presented in Table 2 and Figures 9–13 shows that for the value of the mean time between successive failures amounting to 1000 [h], the reliability value determined in the test in the form of the readiness factor K_g (1) is 0.999815173. The adopted value of the average time between successive failures amounting to 3000 [h] means that the WF devices are fully refurbished and have the property of use at the level of a new facility.

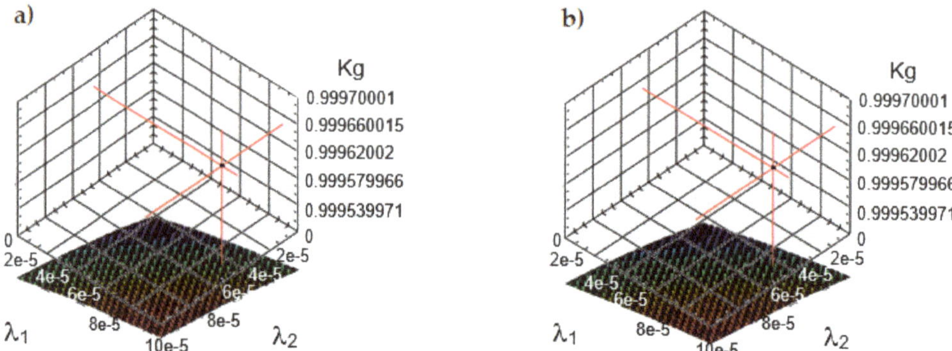

Figure 9. Graphs of the test of the mean time between successive failures, for constant parameters, including: mean repair time 0.7 [h]; mean operating time I type 0.8 [h]; type II mean service time = 1.3 [h], where: (**a**) the mean time between successive failures is 1000 [h]; (**b**) the mean time between successive failures is 1200 [h].

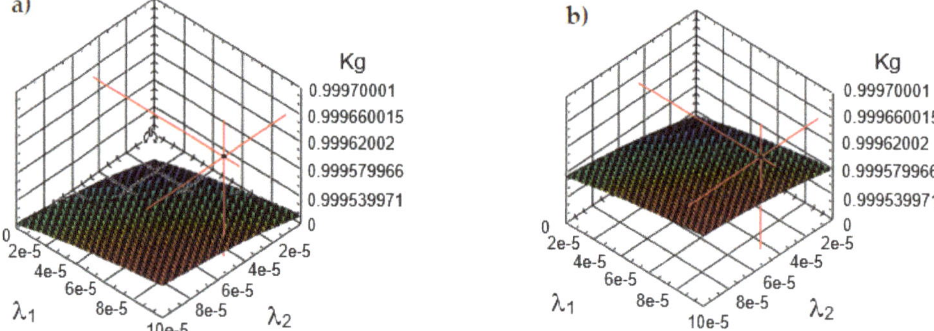

Figure 10. Graphs of the test of the mean time between successive failures, for constant parameters, including: mean repair time 0.7 [h]; mean operating time I type 0.8 [h]; type II mean service time = 1.3 [h], where: (**a**) the mean time between successive failures is 1300 [h]; (**b**) the mean time between successive failures is 1500 [h].

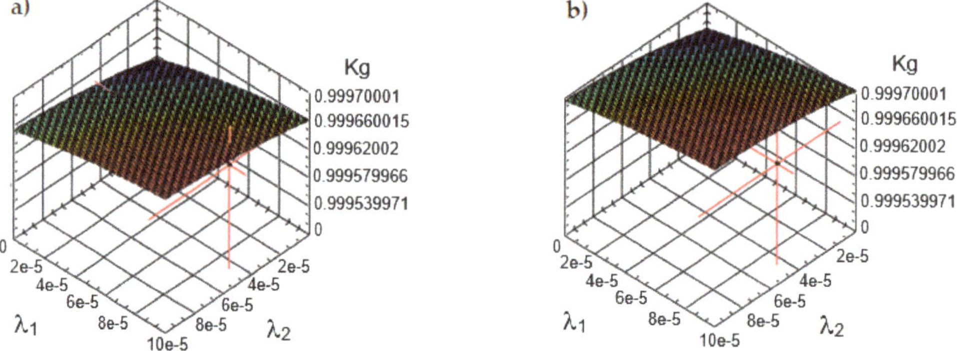

Figure 11. Graphs of the test of the mean time between successive failures, for constant parameters, including: mean repair time 0.7 [h]; mean operating time I type 0.8 [h]; type II mean service time = 1.3 [h], where: (**a**) the mean time between successive failures is 1800 [h]; (**b**) the mean time between failures is 2000 [h].

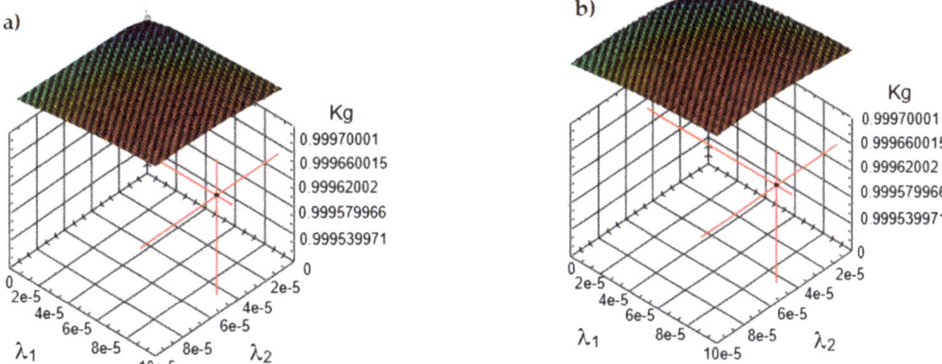

Figure 12. Graphs of the test of the mean time between successive failures, for constant parameters, including: mean repair time 0.7 [h]; mean operating time I type 0.8 [h]); type II mean service time = 1.3 [h], where: (**a**) the mean time between successive failures is 2500 [h]; (**b**) the mean time between successive failures is 3000 [h].

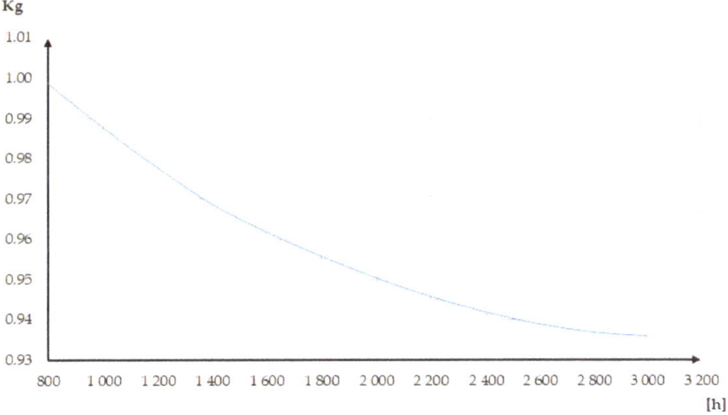

Figure 13. Graph of the reliability of wind farm devices in the operation process as a function of the time between successive failures.

The impact of the average time between successive failures on the quality of the service process, as well as the impact on the level of reliability of WF devices, was examined for this value in the interval {1000 ÷ 3000 [h]} (Figure 13).

The study assumes that the value of the average time between successive failures amounting to 1000 [h] is appropriate for the reliability level of wind farm equipment obtained as a result of the renewal carried out in the maintenance system organized in a traditional (classic) manner [40,41].

The research verified (tested) the flow of the average time between successive failures, amounting to (1000 [h]), appropriate for the renewal of wind farm devices in the mainte-nance system organized classically (without support from the use of intelligent systems). In the meantime between successive failures of (3000 [h]), the reliability determined in the simulation test in the form of the readiness factor K_g (2) of wind farm devices renewed in the service process is 0.99942755.

The analysis of the K_g (1) and K_g (2) readiness factor values allows for determining the possible increase in the value in the form of the readiness factor ΔK_g based on the relationship: $\Delta K_g = K_g$ (1) $- K_g$ (2), hence the value $\Delta K_g = 0.0003876$. The analysis of the research results presented in (Figure 1) allows for the formulation of the following conclusions.

1. Research on the reliability of wind farm devices for the meantime between successive failures confirms that the renewal of wind farm devices in a maintenance system organized on the basis of information developed in intelligent systems, including diagnostic DIAG and maintenance SERV significantly increases the reliability of the renovated facility.
2. The use of efficient and modern intelligent systems for renewing wind farm devices brings a significant increase in the reliability of the renewable wind farm devices in the form of the readiness factor value is $\Delta K_g = 0.0003876$.
3. The developed intelligent SERV system is a good computer tool that effectively supports the design and organization of systems for renewing wind farm devices.

5. Discussion

The problems presented in this paper in reliability studies of the exploitation process of the WF wind farm equipment are particularly well presented in the publication [4]. This paper presents a study of wind farm equipment in the aspect of reliability testing in the aspect of the application of an intelligent expert system to support decision-making by the system operator in the safe supervision of its use. In this work, the following three models of the operation process were developed:

- Model A, reporting the operation process of wind power installations equipped with intelligent systems supporting decision-making regarding the safety in use,
- Model B, describing the process of operating wind farm equipment not equipped with any smart support schemes and without the WPPES system,
- Model C, reporting a simple (conceptual) process of operating wind farm facilities.

For the analytical description of the developed models of the WF equipment operation process, the Kolmogorov–Chapman equations were adopted, which are widely presented in the literature [24]—Nakagawa, T. Maintenance Theory of Reliability; Springer: London, UK, 2005. In the simulation analysis of the reliability of the Wind Farm equipment, the basic reliability quantity of the WF understudy will be determined in the form of the readiness factor $K_g(t)$.

The novelty of this paper about other publications on the reliability of technical devices is the subject of research.

Only in this work, among other publications, was the study of the reliability of wind farm equipment undertaken in the aspect:

1. Testing and evaluation of the reliability of wind farm devices in the operation process due to the decrease in the value of the repair time.
2. Testing and evaluation of the reliability of wind farm devices in the operation process due to changes in the value of time between successive failures of wind farm devices.

The subject matter in the first point of the research is extremely important and of great practical significance. Such research problems as those presented in points 1 and 2 were not presented in publications. The research topic undertaken in pt. 1. concerns modernization of the process of renewing wind farm equipment used in the process of its exploitation. The paper presents that the modernization of the renewal process of the WF is performed by using an intelligent SERV renewal system. Based on information from the SERV system, the service elements are renewed for the appropriate level based on their current state from the {3, 2, 1, 0} set. Then, the implementation of the renewal process (implementation of technical and technological activities) is also optimal. Only those technological activities (renewing service elements) are performed that are required (developed by the SERV system) to fully renew the elements of the WF devices. Therefore, the presented problems of organization and implementation of the wind farm equipment renewal process will result in reducing the repair time.

Having applied the SERV system in the operation process of WF equipment, the task of a simulation study of its effectiveness was undertaken. For the simulation study, it was assumed that decreasing the duration of the process of renewal T_{odn} of WF devices will be studied for the interval {0.2 to 1.00} hour. For such an interval of possible changes in the duration of the process of renewal T_{odn} of WF devices, an increase in the tested value of the readiness factor $\Delta Kg = 0.045$ was obtained. On this basis, it can be concluded that the improvement (modernization) of the process of renewal T_{odn} of WF devices is an important task and affects the increase in the reliability of WF devices.

Another research problem presented in the second point of the research concerned the influence of the value of time between T_{mu} in the operation process of PE equipment on its reliability. It is research that is new in publications on PE topics. In many publications [7,16,27–39] the time between failures T_{mu} in the process of technical equipment operation is identified with the reliability index of these devices. On this basis, it can be said that the greater the value of the time between failures T_{mu} in the process of technical equipment operation, this equipment has higher reliability. In this paper, the task was undertaken to check in practice in research how the value of time between failures T_{mu} in the process of WF equipment operation affects the increase of its reliability. In the conducted research it was assumed that the value of time between failures T_{mu} in the process of operation of WF equipment will vary in the range {from 1000 to 3000} hours. The results obtained from the tests show that the possible growth of the studied magnitude of the readiness factor $\Delta K_g = 0.07$. The value of this reliability measure in the form of a readiness factor K_g is significant. Therefore, it can be concluded that the pursuit, in the process of WF equipment operation, to increase the value of time between failures T_{mu} in the operation process of PE equipment is an important direction for increasing reliability.

6. Conclusions

The problem of testing the reliability properties of wind farm equipment during its operation, as presented in the article, is a difficult organizational and technical task. The difficulty of this result is also due to the acquisition of input data for the research. Numerical data describing the operation process of the Wind Farm equipment was obtained through research carried out over a long period of time. It was assumed that the observation time (measurement of downtime and useful life, etc.) would be sufficient for one year. On the other hand, the reliability tests of the Wind Farm devices will be carried out as a simulation test. This type of research requires the knowledge and description of the actual operation process of the Wind Farm equipment and the determination of reliable input data for the research. At the core of each research is a good research plan (how and how to test) for the wind farm equipment. The basis for the simulation study of the operation process of the Wind Farm equipment is the developed model of the organization of the operation process. Hence, a model of the wind farm equipment operation process was developed, which in the literature is called the four-stage model. The following reliability values were investigated in the simulation tests to determine the reliability of the Wind Farm device in the process of operation with changes:

- the value of the average repair time equal to (0.3) [h], the reliability value in the form of the K_g (1) readiness coefficient determined in the test is the highest and amounts to 0.99979371,
- the value of the mean time between successive failures (3000 [h]), the reliability determined in the simulation test in the form of the readiness factor K_g (2) of wind farm devices renewed in the service process is the highest and amounts to 0.99942755.

Author Contributions: Conceptualization, resources, methodology, software, validation, A.O.; formal analysis, investigation, data curation, A.H. and M.W.; writing—original draft preparation, writing—review and editing, visualization, S.D. and R.D.; supervision, project administration, funding acquisition, J.P. All authors have read and agreed to the published version of the manuscript.

Funding: This research was funded by the Faculty of Electronics and Informatics, Technical University of Koszalin, 2. Sniadeckich St., 75-620 Koszalin, Poland.

Data Availability Statement: The data presented in this article are available at the request of the corresponding author.

Conflicts of Interest: The authors declare no conflict of interest.

Abbreviation

Symbols

$X(e_{i,j})$	diagnostic signal in jth element of ith set
$X_{(w)}(e_{i,j})$	model signal for $X(e_{i,j})$ signal
$F_{C\,max}$	max. value of the function of the use of the object
$W(\varepsilon(e_{i,j})) = \{2, 1, 0\})$	valued of state assessment logics for jth element within ith module (from the set of the accepted three-value logic of states' assessment)
$K_g(t)$ or K_g	the average value of availability function or factor K_g
F_c	the quality function of the object's operation process
F_{ch}	function of the object exploitation process
λ	damage intensity
T_o	simulation test time of the object
μ	repair intensity
λ_1	intensity of type I inspections
μ_1	type I operational maintenance intensity
λ_2	intensity of type II inspections
μ_2	type II operational maintenance intensity
P_0	probability of the system being in state S0
P_1	probability of the system being in state S1
P_{01}	probability of the system being in state S01
P_{10}	probability of the system being in state S10
S0	effective use of the object
S1	scheduled maintenance–preventive NP
S01	unscheduled maintenance
S10	repair, restoration of utility properties ineffective use of the object

Acronyms

WPPES	Wind Power Plant Expert System
SERV	intelligent operating system
DIAG	intelligent diagnostic system

References

1. Abo-Khalil, A.G.; Alghamdi, A.I.; Tlili, A.; Eltamaly, A.M. Current Controller Design for DFIG-based 426 Wind Turbines Using State Feedback Control. *IET Renew. Power Gener.* **2019**, *13*, 1938–1949. [CrossRef]
2. Eltamaly, A.M. Modeling of wind turbine driving permanent magnet generator with maximum power 383point tracking system. *J. King Saud Univ.-Eng. Sci.* **2007**, *19*, 223–236.
3. Andalib, C.; Liang, X.; Zhang, H. Fuzzy-Secondary-Controller-Based Virtual Synchronous Generator 386 Control Scheme for Interfacing Inverters of Renewable Distributed Generation in Microgrids. *IEEE Trans. Ind. Appl.* **2018**, *54*, 1047–1061. [CrossRef]
4. Duer, S. Assessment of the Operation Process of Wind Power Plant's Equipment with the Use of an Artificial Neural Network. *Energies* **2020**, *13*, 2437. [CrossRef]
5. Badrzadeh, B.; Gupta, M.; Singh, N.; Petersson, A.; Max, L.; Høgdahl, M. Power system harmonic analysis in wind power plants-Part I: Study methodology and techniques. In Proceedings of the IEEE Industry Applications Society Annual Meeting, Las Vegas, NV, USA, 7–11 October 2012; pp. 1–11.
6. Kunjumuhammed, L.P.; Pal, B.C.; Oates, C.; Dyke, K.L. Electrical oscillations in wind farm systems: Analysis and insight based on detailed modeling. *IEEE Trans. Sustain. Energy* **2016**, *7*, 51–62. [CrossRef]

7. Pogaku, N.; Prodanovic, M.; Green, T.C. Modeling, analysis and testing of autonomous operation of an inverter-based microgrid. *IEEE Trans. Power Electron.* **2007**, *22*, 613–625. [CrossRef]

8. Shahanaghi, K.; Babaei, H.; Bakhsha, A. A Chance Constrained Model for a Two Units Series Critical System Suffering from Continuous Deterioration. *Int. J. Ind. Eng. Prod. Res.* **2009**, *20*, 69–75.

9. Sun, J. Impedance-based stability criterion for grid-connected inverters. *IEEE Trans. Power Electron.* **2011**, *26*, 3075–3078. [CrossRef]

10. Hojjat, A.; Shih, L.H. *Machine Learning, Neural Networks, Genetic Algorithms and Fuzzy Systems*; John Wiley & Sons, Inc.: Hoboken, NJ, USA, 1995; p. 398.

11. Buchannan, B.; Shortliffe, E. *Rule—Based Expert Systems*; Addison—Wesley Publishing Company: London, UK; Amsterdam, The Netherlands; Don Mills, ON, Canada; Sydney, Australia, 1985; p. 387.

12. Hayer-Roth, F.; Waterman, D.; Lenat, D. *Building Expert Systems*; Addison—Wesley Publishing Company: Boston, MA, USA, 1983; p. 321.

13. Waterman, D. *A Guide to Export Systems*; Addison—Wesley Publishing Company: Boston, MA, USA, 1986.

14. Wiliams, J.M.; Zipser, D. A learning Algorithm for Continually Running Fully Recurrent Neural Networks. *Neural Comput.* **1989**, *1*, 270–280. [CrossRef]

15. Linz, P. *An Introduction to Formal Languages and Automata*; University of California: Davis, CA, USA, 2002.

16. Bernatowicz, D.; Duer, S.; Wrzesień, P. Expert system supporting the diagnosis of the wind farm equipments. In *Communications in Computer and Information Science*; Springer: Poznan, Poland, 2018; Volume 928, pp. 432–441.

17. Gupta, M.M.; Jin, L.; Homma, N. *Static and Dynamic Neural Networks, From Fundamentals to Advanced Theory*; John Wiley and Sons, Inc.: Hoboken, NJ, USA, 2003; p. 718.

18. Tang, L.; Liu, J.; Rong, A.; Yang, Z. Modeling and genetic algorithm solution for the slab stack shuffling problem when implementing steel rolling schedules. *Int. J. Prod. Res.* **2002**, *40*, 272–276. [CrossRef]

19. Mathirajan, M.; Chandru, V.; Sivakumar, A.I. Heuristic algorithms for scheduling heat-treatment furnaces of steel casting industries. *Sadahana* **2007**, *32*, 111–119. [CrossRef]

20. Kacalak, W.; Majewski, M. New Intelligent Interactive Automated Systems for Design of Machine Elements and Assemblies. In *Lecture Notes in Computer Science*; Springer: Berlin, Germany, 2012; pp. 115–122.

21. Lipinski, D.; Majewski, M. System for monitoring and optimization of micro- and nano-machining processes using intelligent voice and visual communication. In *Lecture Notes in Computer Science*; Springer: Berlin, Germany, 2013; Volume 8206, pp. 16–23.

22. Majewski, M.; Kacalak, W. Smart control of lifting devices using patterns and antipatterns. In *Artificial Intelligence Trends in Intelligent Systems*; Advances in intelligent systems and computing; Springer: Cham, Switzerland, 2017; Volume 573, pp. 486–493. [CrossRef]

23. Majewski, M.; Kacalak, W. Innovative intelligent interaction systems of loader cranes and their human operators. In *Artificial Intelligence Trends in Intelligent Systems*; Advances in intelligent systems and computing; Springer: Cham, Switzerland, 2017; Volume 573, pp. 474–485. [CrossRef]

24. Duer, S.; Bernatowicz, D.; Wrzesień, P.; Duer, R. The diagnostic system with an artificial neural network for identifying states in multi-valued logic of a device wind power. In *Communications in Computer and Information Science*; Springer: Poznan, Poland, 2018; Volume 928, pp. 442–454.

25. Zurada, I.M. *Introduction to Artificial Neural Systems*; West Publishing Company: St. Paul, MN, USA, 1992; p. 324.

26. Bedkowski, L.; Dabrowski, T. *Basic of the Maintenance Theory*; Publishing House of WAT: Warsaw, Poland, 2006; p. 187.

27. Dyduch, J.; Paś, J.; Rosiński, A. *The Basic of the Exploitation of Transport Electronic Systems*; Publishing House of Radom University of Technology: Radom, Poland, 2011.

28. Epstein, B.; Weissman, I. *Mathematical Models for Systems Reliability*; CRC Press/Taylor & Francis Group: Boca Raton, FL, USA, 2008.

29. Klimczak, T.; Paś, J. *Selected Issues of the Reliability and Operational Assessment of a Fire Alarm System*; Maintenance and Reliability: Warsaw, Poland, 2019; Volume 21, pp. 553–561.

30. Siergiejczyk, M.; Paś, J.; Rosiński, A. Issue of reliability–exploitation evaluation of electronic transport systems used in the railway environment with consideration of electromagnetic interference. *IET Intell. Transp. Syst.* **2016**, *10*, 587–593. [CrossRef]

31. Siergiejczyk, M.; Rosiński, A. Analysis of power supply maintenance in transport telematics system. *Solid State Phenom.* **2014**, *210*, 14–19. [CrossRef]

32. Dhillon, B.S. *Applied Reliability and Quality, Fundamentals, Methods and Procedures*; Springer: London, UK, 2006; p. 186.

33. Rychlicki, M.; Kasprzyk, Z.; Rosiński, A. Analysis of Accuracy and Reliability of Different Types of GPS Receivers. *Sensors* **2020**, *20*, 6498. [CrossRef] [PubMed]

34. Stawowy, M.; Olchowik, W.; Rosiński, A.; Dąbrowski, T. The Analysis and Modelling of the Quality of Information Acquired from Weather Station Sensors. *Remote Sens.* **2021**, *13*, 693. [CrossRef]

35. Paś, J.; Rosiński, A.; Chrzan, M.; Białek, K. Reliability-Operational Analysis of the LED Lighting Module Including Electromagnetic Interference. *IEEE Trans. Electromagn. Compat.* **2020**, *62*, 2747–2758. [CrossRef]

36. Nakagawa, T. *Maintenance Theory of Reliability*; Springer: London, UK, 2005.

37. Nakagawa, T.; Ito, K. Optimal inspection policies for a storage system with degradation at periodic tests. *Math. Comput. Model.* **2000**, *31*, 191–195.

38. Pokoradi, L. Logical Tree of Mathematical Modeling. *Theory Appl. Math. Comput. Sci.* **2015**, *5*, 20–28.

39. Dempster, A.P. Upper and lower probabilities inducted by a multi-valued mapping. *Ann. Math. Stat.* **1967**, *38*, 325–339. [CrossRef]
40. Smyczek, J.; Zajkowski, K. Simulation of overvoltages for switching off lagging load from mains. In Proceedings of the 2nd International Industrial Simulation Conference, Malaga, Spain, 7–9 June 2004; pp. 278–281.
41. Zajkowski, K. Settlement of reactive power compensation in the light of white certificates. In *E3S Web of Conferences 19, UNSP 01037*; EDP Sciences: Les Ulis, France, 2017. [CrossRef]
42. Zajkowski, K. The method of solution of equations with coefficients that contain measurement errors, using artificial neural network. *Neural Comput. Appl.* **2012**, *24*, 431–439. [CrossRef]
43. Zajkowski, K. An innovative hybrid insulation switch to enable/disable electrical loads without over voltages. In *E3S Web of Conferences 19, UNSP 01033*; EDP Sciences: Les Ulis, France, 2017. [CrossRef]
44. Zajkowski, K. Two-stage reactive compensation in a three-phase four-wire systems at no sinusoidal periodic waveforms. *Electr. Power Syst. Res.* **2020**, *184*, 106296. [CrossRef]
45. Duer, S.; Zajkowski, K.; Harničárová, M.; Charun, H.; Bernatowicz, D. Examination of Multivalent Diagnoses Developed by a Diagnostic Program with an Artificial Neural Network for Devices in the Electric Hybrid Power Supply System "House on Water". *Energies* **2021**, *14*, 2153. [CrossRef]

MDPI
St. Alban-Anlage 66
4052 Basel
Switzerland
Tel. +41 61 683 77 34
Fax +41 61 302 89 18
www.mdpi.com

Energies Editorial Office
E-mail: energies@mdpi.com
www.mdpi.com/journal/energies